John Wesley Powell

Truth and Error

Or the Science of Intellection

John Wesley Powell

Truth and Error
Or the Science of Intellection

ISBN/EAN: 9783337035501

Printed in Europe, USA, Canada, Australia, Japan

Cover: Foto ©berggeist007 / pixelio.de

More available books at **www.hansebooks.com**

TRUTH AND ERROR

TRUTH AND ERROR

OR

THE SCIENCE OF INTELLECTION

BY

J. W. POWELL

CHICAGO
THE OPEN COURT PUBLISHING COMPANY
(LONDON: KEGAN PAUL, TRENCH, TRÜBNER & CO.)
1898

TO
LESTER F. WARD
PHILOSOPHER AND FRIEND, I DEDICATE

THIS BOOK

CONTENTS

CHAP.		PAGE
I.	Chuar's Illusion	1
II.	Essentials of Properties	9
III.	Quantities or Properties that are Measured	20
IV.	Kinds or Properties that are Classified	31
V.	Processes or the Properties of Geonomic Bodies	42
VI.	Generations or Properties of Plants	64
VII.	Principles or Properties of Animals	74
VIII.	Qualities	98
IX.	Classification	109
X.	Homology	133
XI.	Dynamics	152
XII.	Coöperation	168
XIII.	Evolution	183
XIV.	Sensation	207
XV.	Perception	226
XVI.	Apprehension	237
XVII.	Reflection	251
XVIII.	Ideation	264
XIX.	Intellections	278
XX.	Fallacies of Sensation	307
XXI.	Fallacies of Perception	335
XXII.	Fallacies of Apprehension	352
XXIII.	Fallacies of Reflection	374
XXIV.	Fallacies of Ideation	391
XXV.	Summary	413
	Index	425

TRUTH AND ERROR

CHAPTER I

CHUAR'S ILLUSION

IN the fall of 1880 I was encamped on the Kaibab plateau above the canyon gorge of a little stream. White men and Indians composed the party with me. Our task was to make a trail down this side canyon, which was many hundreds of feet in depth, into the depths of the Grand Canyon of the Colorado. While in camp after the day's work was done, both Indians and white men amused themselves by attempting to throw stones across the little canyon. The distance from the brink of the wall on which we were encamped to the brink of the opposite wall seemed not very great, yet no man could throw a stone across the chasm, though Chuar, the Indian chief, could strike the opposite wall very near its brink. The stones thrown by others fell into the depths of the canyon. I discussed these feats with Chuar, leading him to an explanation of gravity. Now Chuar believed that he could throw a stone much farther along the level of the plateau than over the canyon. His first illusion was thus one very common among mountain travelers—an underestimate of the distance of towering and massive rocks when the eye has no intervening objects to divide the space into parts as measures of the whole.

I did not venture to correct Chuar's judgment, but simply sought to discover his method of reasoning. As our conversation proceeded he explained to me that the stone could not go far over the canyon, for it was so deep that it would make the stone fall before reaching the opposite bank; and he explained to me with great care that the hollow or empty space pulled the stone down. He discoursed on this point at length, and illustrated it in many ways: "If you stand on the edge of the cliff you are likely to fall; the hollow pulls you down, so that you are compelled to brace yourself against the force and lean back. Any one can make such an experiment and see that the void pulls him down. If you climb a tree the higher you reach the harder the pull; if you are at the very top of a tall pine you must cling with your might lest the void below pull you off."

Thus my dusky philosopher interpreted a subjective fear of falling as an objective force; but more, he reified void and imputed to it the force of pull. I afterward found these ideas common among other wise men of the dusky race, and once held a similar conversation with an Indian of the Wintun on Mount Shasta, the sheen of whose snow-clad summit seems almost to merge into the firmament. On these dizzy heights my Wintun friend expounded the same philosophy of gravity.

Now, in the language of Chuar's people, a wise man is said to be a traveler, for such is the metaphor by which they express great wisdom, as they suppose that a man must learn by journeying much. So in the moonlight of the last evening's sojourn in the camp on the brink of the canyon, I told Chuar that he was a great traveler, and that I knew of two other great travelers among the seers of the East, one by the name of Hegel, and another by the name of Spencer, and that I should ever remember these three wise men, who spoke like words of wisdom, for it passed through my mind that all three of these philosophers had reified void and founded a philosophy thereon.

Concepts of number, space, motion, timé and judgment are developed by all minds, from that of the lowest animal to that of the highest human genius. Through the evolution of animal life, these concepts have been growing as they have been inherited down the stream of time in the flood of generations. It is thus that an experience has been developed, combined with the experience of all the generations of life for all the time of life, which makes it impossible to expunge from human mind these five concepts. They can never be canceled while sanity remains. Things having something more than number, space, motion, time and judgment cannot even be invented; it is not possible for the human mind to conceive anything else, but semblances of such ideas may be produced by the mummification of language.

Ideas are expressed in words which are symbols, and the word may be divested of all meaning in terms of number, space, motion, time and judgment and still remain, and it may be claimed that it still means something unknown and unknowable; this is the origin of reification. There are many things unknown at one stage of experience which are known at another, so man comes to believe in the unknown by constant daily experience; but has by further converse with the universe known things previously unknown, and they invariably become known in terms of number, space, motion, time and judgment, and are found to be only combinations of these things. It is thus that something unknown may be conceived, but something unknowable cannot be conceived.

No man conceives reified substrate, reified essence, reified space, reified force, reified time, reified spirit. Words are blank checks on the bank of thought, to be filled with meaning by the past and future earnings of the intellect. But these words are coin signs of the unknowable and no one can acquire the currency for which they call.

Things little known are named and man speculates about these little-known things, and erroneously imputes properties or attributes to them until he comes to think of them as possessing such unknown and mistaken attributes. At last he discovers the facts; then all that he discovers is expressed in the terms of number, space, motion, time and judgment. Still the word for the little-known thing may remain to express something unknown and mystical, and by simple and easily understood processes he reifies what is not, and reasons in terms which have no meaning as used by him. Terms thus used without meaning are terms of reification.

Such terms and such methods of reasoning become very dear to those immersed in thaumaturgy and who love the wonderful and cling to the mysterious, and, in the revelry developed by the hashish of mystery, the pure water of truth is insipid. The dream of intellectual intoxication seems more real and more worthy of the human mind than the simple truths discovered by science. There is a fascination in mystery and there has ever been a school of intellects delighting to revel therein, and yet, in the grand aggregate, there is a spirit of sanity extant among mankind which loves the true and simple.

Often the eloquence of the dreamer has even subverted the sanity of science, and clear-headed, simple-minded scientific men have been willing to affirm that science deals with trivialities, and that only metaphysics deals with the profound and significant things of the universe. In a late great text-book on physics, which is a science of simple certitudes, it is affirmed:

"To us the question, *What is matter?*—What is, assuming it to have a real existence outside ourselves, the essential basis of the phenomena with which we may as physicists make ourselves acquainted?—appears absolutely insoluble. Even if we become perfectly and certainly acquainted with the intimate structure of what we call Matter, we would but have made a further step in the study of its properties; and as physicists we are forced to say that while somewhat has been learned as to the properties of Matter, its essential nature is quite unknown to us."

As though its properties did not constitute its essential nature.

So, under the spell of metaphysics, the physicist turns from his spectroscope to exclaim that all his researches may be dealing with phantasms.

Science deals with realities. These are bodies with their properties. All the facts embraced in this vast field of research are expressed in terms of number, space, motion, time and judgment; no other terms are needed and no other terms are coined, but by a process well known in philology as a disease of language, sometimes these terms lapse into meanings which connote fallacies. The human intellect is of such a nature that it has notions or ideas which may be certitudes or fallacies. All the processes of reasoning, including sensation and perception, proceed by inference; the inference may be correct or erroneous, and certitudes are reached by verifying opinions. This is the sole and only process of gaining certitudes. The certitudes are truths which properly represent noumena, the illusions are errors which misrepresent noumena. All knowledge is the knowledge of noumena, and all illusion is erroneous opinion about noumena. The human mind knows nothing but realities and deals with nothing but realities, but in this dealing with the realities—the noumena of the universe—it reaches some conclusions that are correct and others that are incorrect. The correct conclusions are certitudes about realities; the incorrect conclusions are fallacies about realities. ' Science is the name which mankind has agreed to call this knowledge of realities, and error is the name which mankind has agreed to give to all fallacies. Thus it is that certitudes are directly founded upon realities; and fallacies alike all refer to realities. In this sense then it may be stated that all error as well as knowledge testifies to reality, and that all our knowledge is certitude based upon reality, and that fallacies would not be possible were there not realities about which inferences are made.

Known realities are those about which mankind has knowledge; unknown things are those things about which man has not yet attained knowledge. Scientific research is the endeavor to increase knowledge, and its methods are experience, observation and verification. Fallacies are erroneous inferences in relation to known things. All certitudes are described in terms of number, space, motion, time and judgment; nothing else has yet been discovered and nothing else can be discovered with the faculties with which man is possessed.

In the material world we have no knowledge of something which is not a unity of itself or a unity of a plurality; of something which is not an extension of figure or an extension of figure and structure; of something which has not motion or a combination of motions as force; of something which has not duration as persistence or duration with persistence and change.

In the mental world we have no knowledge of something which is not a judgment of consciousness and inference; of a judgment which is not a judgment of a body with number, space, motion and time. Every notion of something in the material world devoid of one or more of the constituents of matter is an illusion; every notion of something in the spiritual world devoid of the factors of matter and judgment is a fallacy. These are the propositions to be explained and demonstrated.

In the following chapters an attempt will be made to show that we know much about matter, and although we do not know all, all we know is about matter in its essentials of number, space, motion, time and judgment, or that we know of matter in its four essentials and of mind as consciousness exhibited in judgment and concepts, but always this mind is associated with matter. In doing this we shall endeavor to discriminate between the certitudes and fallacies current in human opinion.

In the intoxication of illusion facts seem cold and colorless, and the wrapt dreamer imagines that he dwells in a realm above science—in a world which as he thinks absorbs truth as the ocean the shower, and transforms it into a flood of philosophy. Feverish dreams are supposed to be glimpses of the unknown and unknowable, and the highest and dearest aspiration is to be absorbed in this sea of speculation. Nothing is worthy of contemplation but the mysterious. Yet the simple and the true remain. The history of science is the history of the discovery of the simple and the true; in its progress fallacies are dispelled and certitudes remain.

CHAPTER II

ESSENTIALS OF PROPERTIES

On the threshold it is necessary to state certain scientific conclusions which I accept. These are the four great doctrines taught by modern science. I accept the atomic theory that the constitution of bodies is explained as a numerical combination of ultimate smaller particles. I accept the modern doctrine of morphology, that forms in different kinds of bodies exhibit homologies that express degrees of relationship. I accept the modern doctrine of the persistence of motion as the proper explanation of the correlation of forces. I accept the modern doctrine of evolution, that higher bodies are derived from lower. In accepting these doctrines I try to embrace them in all their logical results, some of which may seem strange to my readers. I shall propound the hypothesis that consciousness inheres in the ultimate particle, and attempt to show that it harmonizes the principles of psychology.

The four great doctrines of modern science which I have enumerated were originally guesses, but they have largely been accepted by scientific men because they explain the phenomena of the universe to which they relate. The chaos of scientific phenomena collected in vast catalogues of facts are seen to be explained by these laws.

The chemical theory may be denominated the persistence of units; the morphologic theory the persistence of extensions; the dynamic theory the persistence of speeds; the evolutionary theory the persistence of existence.

There are systems of stars, and every system is a body. The one to which our earth belongs is well known, for the solar system is the theme of the venerable science of astronomy. The earth itself is composed of four grand bodies: an outer envelope of air or atmosphere, a middle envelope of water or hydrosphere, an inner envelope of rock or lithosphere, and the grand central nucleus or centrosphere. Neglecting the two outer envelopes and considering only the stony crust, we find that it is composed of many bodies or formations and these of rocks, while there are many plants and animals, and all again are divided into grains, crystals or cells, and the grains, crystals or cells are divided into molecules, and molecules are composed of other molecules, until at last chemical atoms are reached; so it is discovered that the universe is a hierarchy of bodies.

The universe is a hierarchy of bodies composed of bodies and these again composed of bodies in a vast succession as they are reduced by analysis. When we come to discuss the relations of these bodies to one another it will be convenient and conduce to exact expression if we make a distinction between bodies and particles, and speak of a body when we wish to consider it as a unit and then speak of its particles when we wish to speak of the parts of which it is composed. A body, therefore, is a body of particles which are many in one, the one being a body; the many particles severally may be bodies composed of particles, that is, one composed of many. The solar system is a body of particles, the particles being the stars of which it is composed; the earth, one of these particles, may be considered as a body, when its particles will be the air, the water, the stony crust and the central nucleus; then the air may be considered as a body composed of many particles, the water may be considered as a body composed of particles, the stony crust as a body composed of particles, and finally the nucleus as a body composed of particles. In this sense it will be understood we sometimes speak of something as a body and again of the same thing as a particle. A body and its particles are reciprocal. When we consider a body as composed of particles we consider internal relations, but when we consider the particles severally their relations to one another are external. Thus a body has internal relations and external relations, and every particle of the body also has internal relations and external relations, if it is composed of parts.

A substance is an aggregation of like particles in one body or a number of bodies. Bodies are composed of substances. For example, the air is a substance which is again composed of substances; the water is a substance, and this water is oxygen and hydrogen and contains in solution many other substances. In the envelope of rock a great variety of substances are discovered; then there are vegetal and animal substances. Thus in the hierarchy of bodies there is discovered to be a hierarchy of substances, extending from elements to protoplasm. The vast multitude of substances have so far been resolved into about seventy seemingly simple substances, but there is reason to believe that they are to be still further resolved into one primordial substance, which is called matter. Matter, then, is the ultimate substance into which all other substances which constitute the bodies of the universe are resolved; and matter may be of one primordial kind, or it may be of seventy kinds, more or less.

Bodies are resolved into more and more simple and homogeneous substances, and it is the theory of some chemists that ultimate analysis will resolve them into one simple kind, so that every particle will be like every other particle in all its properties. Matter, then, is the ultimate kind of particle into which all bodies may be analyzed, and different kinds of matter are different aggregations of the one kind. The different kinds of matter made different by different aggregation are different substances, and the different substances are aggregations of matter by incorporation.

An army is composed of men, but there are platoons, companies, battalions, regiments, divisions, and corps in the army. So it is organized or incorporated into a hierarchy of units. The platoon is one as a platoon, composed of a plurality of men; the company is one as a company but a plurality of platoons; the battalion is one as a battalion but a plurality of companies; the regiment is one as a regiment but a plurality of battalions; the brigade is one as a brigade but a plurality of regiments; the division is one as a division but a plurality of brigades; the corps is one as a corps but a plurality of divisions. Now we understand the fundamental property of numbers as many in one. The platoon differs in the property of number from the individual; the company differs in the property of number from the platoon, and the battalion differs in the property of number from the company; and the same is true of all the units in the hierarchy.

These units of different orders have different properties of space; the platoon occupies more space in the field than the individual soldier; the company occupies more space than the platoon; the battalion more space than the company; and the same is true of the other units in the hierarchy. If we have two armies exactly alike in a hierarchy of units and spaces, then any two corresponding units and spaces in the hierarchy would be similar. In speaking of the bodies of the universe it is necessary sometimes to speak of the corresponding unit in the different bodies, and we call them substances. The oxygen in one molecule of water is the same in all molecules of water, and we call all units a substance. Every body of water is composed of molecules of water, and there are many bodies of water, and we call bodies of water a substance. We thus designate as one substance all like units of matter.

This is very simple. It is merely a statement of the resolution of more compound bodies into simpler bodies and of more compound substances into simpler substances. It is the dissection of bodies in parts and the analysis of substances into elements.

The ultimate particle found in any substance may be still further resolved in consideration. Every body, whether it be a stellar system or an atom of hydrogen, has certain fundamental characteristics found in all. These are number, space, motion and time, and if it be an animate body, judgment. They shall here be known as properties, and to them attention must now be turned.

Let us first consider with what things one inanimate

particle is endowed. First, it must have unity. There must be one, or it does not exist. Second, it must have extension, for without extension it does not exist. Third, it must have speed, for it cannot have motion without speed, nor can it have force without motion, and a particle of matter not in motion is unknown. The body lying upon the ground at rest is not without motion, for it has the motion of the earth about its axis and the motion of the earth about the sun; it also has a motion of its molecules and atoms, which is heat and structural motion. If the body which is lying upon the ground is moved the motions are deflected and it is impossible to discover that any motion as speed is added to them. Rest is only the absence of molar motion. Fourth, the same particle of matter must have persistence, for persistence is necessary to its existence. Here persistence is used to mean continued existence.

I shall attempt to demonstrate the proposition that every particle of matter has consciousness, and hence the fifth property here called judgment, but shall reserve the discussion of the subject to a later part of the work.

One ultimate particle must have essentials that it may exist, but they are all comprehended in one particle. If we consider the essentials separately we call it abstraction; if we consider them conjointly we call it comprehension, and the terms abstraction and comprehension will be used in these senses only.

These essentials are simple and wholly unlike one another. There is nothing in unity like extension, nothing in extension like speed, nothing in speed like persistence. There is no possible way of deriving one from another. We cannot derive extension from unity, but extension must be concomitant with unity; extension and unity are concomitant in one particle. We cannot derive speed from extension, but the thing which has speed must have extension. We cannot derive persistence from speed, but that which has persistence must have speed. So we may run through all permutations of these essentials and find them wholly unlike one another and discover no possible way of deriving one from the other. Notwithstanding their total unlikeness, they are never dissociated so that one exists without the other; they may be considered separately but cannot exist separately. They cannot be analyzed and the unity placed in one box, the extension in a second, the speed in a third, the persistence in a fourth; but they may be considered separately, and this is abstraction as distinguished from analysis. Bodies may be dissected, substances may be analyzed, essentials may be abstracted in consideration.

The essentials are indissoluble in every particle. Where there is no unit there is no extension, no speed and no persistence. Where there is no speed there is no unit, no extension, no persistence. Where there is no persistence there is no unit, no extension and no speed. If any of the essentials of a particle of inanimate matter be taken away, the matter disappears. A particle is the essentials of which it is composed, and it has no other substrate. It exists in its essentials, and its essentials exist in it, and neither existence is separate. The notion of a particle of matter as a substrate of essentials, or as something to which the essentials adhere or inhere and from which they may be taken away, leaving behind the particle, which is not a unit, an extension, a speed and a persistence, is a pseud-idea, the result of mythologizing, where the word is taken to represent more than the sum of the essentials of the object to which it is applied. A unit is a unit of an extension, a speed and a persistence. An extension is an extension of a unit, a speed and a persistence. A speed is a speed of a unit, an extension and a persistence. A persistence is a persistence of a unit, an extension and a speed.

Think of properties as number, space, motion and time; then consider the things which must exist if these properties exist and you have the essentials, as the term is here used. Thus think not of number, but of unity; think not of space, but of extension; think not of motion, but of speed; think not of time, but of persistence, and you have the essentials themselves.

This chapter is designed to define the essentials of an inanimate particle, and to show in what sense the terms for the essentials are used. The mathematician might say that A stands for unity, B for extension, C for speed, D for persistence, E for consciousness, and you would not find fault. Should he formulate an equation you would not quarrel with him about his symbols, because he uses A for apples, B for bushels, C for cents, D for division, and E for equality to show the equity of a transaction represented by F. Let me use my symbols in my manner, if you would understand my demonstration. Unity means one, extension means exclusive occupancy of space, speed means change of position, persistence means continuance in time.

The statement might be left to stand by itself, yet I think it best to explain why I use these terms. About the term unity no one will cavil.

For extension the term impenetrability has been used, but it has a negative connotation which I wish to avoid. I once thought of using dimension, but I soon found that I must use it in another sense in discussing measure. Then I thought of space. Now, space has a metaphysical use in which it is synonymous with vacuum or void and from which I wish to rescue it. So I concluded to use the term extension to signify exclusive occupancy of space, and to use space itself for the extension of positions of extensions, which also includes the extension of the medium which makes up the space. Let this be made clear. As the terms are here used the particles of the walls of this box have extension, and the particles of air which it contains have extensions, and the particles of ether within the air have extensions, but the space of the box includes the extensions of the box, the extensions of the air, and also the extension of the ether. I may speak of the space of the box and refer only to the position of the particles of the box and I may then speak of the space of the box as the sum of the extensions of the walls, air, and ether. It may be that the walls of the box have minute apertures in which air exists, so that all the air is not excluded from the wood, and it is certain that the ether is not excluded from the wood. And it may be that there are interspaces between the particles of wood, air and ether. Therefore even the wood of the box must be described in terms of space, not in terms of extension. When we come to discuss extension itself, we find ourselves considering mass, so that mass and extension are here nearly synonymous; but mass is used as the measure of extensions, while space is the dimensions of related positions. Mass is the measure of the numbers of particles of extension, but units of space are measured with units of length.

I use the term speed because in modern physics it has exactly the meaning which I desire. The popular meaning of velocity is just what I need, but in physics velocity means rate of speed and also rate of deflection and the term is needed for that purpose.

I use the term persistence because the term time or the term duration means persistence and change or they may mean the measure of states separated by change, while the term persistence is free from these implications.

If the terms are understood we are ready to proceed to another stage of exposition.

Essentials are comprehended in the same particle, and we shall call them concomitants. We shall not say that one essential is related to another in the same particle, but they are concomitant with one another, though the essential of one particle may be related to the essential of another particle. A unit may be related to another unit, an extension may be related to another extension; but the unit and the extension in the same particle are not related to each other but concomitant with each other, and these same distinctions must be observed with all the essentials. The task before us in this chapter is the exhibition of the concomitants of particles and relations of essentials, concomitants inhering in every particle, the relations arising by reason of the relation of particles to particles.

The student who follows my argument must first become accustomed to the discrimination between concomitancy and relativity. Relativity is the relation of one particle or body to another; concomitancy is the coexistence of one property with another in the same particle or body.

Having deduced or discovered four essentials or concomitants in every particle of matter, we have yet to determine whether these are all, and for this purpose we are compelled to assemble in a passing review all of the bodies of the universe. To do this it becomes necessary to discover in what manner these four essentials become properties as quantities and kinds, for we have quantitative properties and classific properties. Having discovered how the essentials become properties, we can then go on in the review of the universe of bodies.

CHAPTER III

QUANTITIES OR PROPERTIES THAT ARE MEASURED

Two short chapters must now be presented which will be found rather dry, but they must be mastered if the subsequent chapters are to be understood. The principles therein stated are the A, B, C, of the work—the multiplication table of our logic. I beg of my reader not to be deterred from their careful consideration by reason of their simplicity.

I

The universe is a concourse of related bodies composed of related particles. Every relation must exist between two or more particles or bodies, and every particle or body is related to every other particle or body directly or indirectly. The universe is a hierarchy of bodies, and thus there is a hierarchy of relations. A relation cannot exist independent of terms. We may consider a relation abstractly, but it cannot exist abstractly. To affirm a relation the terms must be implied. When an abstract is reified, that is, supposed to exist by itself independent of other essentials, and the illusion is entertained that there is something independent of the essentials which support them, a mythology is created so subtle as to simulate reality. So when relations are reified and supposed to exist independent of terms, the mind is astray in the realm of fallacies. When it is discovered that rest is only a relation, the mind is prone to believe that nothing exists but relation, for we have often discovered that which we thought was absolute was in fact relation; but rest is a relation between terms which are absolute. The internal or molecular motions of the body at rest have a certain relation to the external or astronomic motions of the body which are changed when the body is given molar motion, but the absolutes still remain, though deflected.

Human beings are molar bodies, and have a deep interest in one another as such and in the other molar bodies with which they are associated. Molar bodies and their relations are the first bodies discovered by primitive man, and his converse with the external world at first seems to be wholly with molar bodies. Molar bodies are those in which he first discovers relations and with which he first consciously and purposely associates, and they become the type of the others. Molecular bodies are known as such only to science. The stellar bodies are first believed to be molar bodies, and it is long before the corporeal structure of the earth is discovered as a body of great magnitude associated with other bodies more nearly commensurate with them, as the sun, moon and stars.

Of the internal relations of molecular bodies little is known even yet, and in the same manner of the internal relations of stellar bodies, but little is yet known. Our ideas of molecular and stellar bodies are largely ideas of their individuality, or as units related to units of the same order, while their constituent units scarcely receive consideration. In the mechanical or molar world the relations of parts are immeasurably more numerous than the parts themselves. Not only are rocks multifarious and the imperfect embodiments of air and water multifarious, but special classes of embodiments are discovered as plants and animals distributed over all the earth in multitudinous kinds with multitudinous relations, and men as molar bodies are related to one another and in all of these relations men are fundamentally interested.

Relations, therefore, are so great in number and so many in kind that the subject of relations is apt to overwhelm the mental powers, for man discovers that in his reasoning he is forever dealing with relations far more than directly with the bodies themselves. In this manner he discovers that the world is a congress of molar bodies that are related to one another through their properties; when they are analyzed into related particles or synthesized into related bodies, relation seems to swallow all else, so that philosophers often assume and sometimes affirm that all that is known of the universe is these relations, and finally that the universe is only a system of relations and the substantiality of the universe is denied. The universe thus becomes a universe of relations without terms. The confounding of concomitancy with relativity is a cause of inextricable confusion—a snare to the intellect and a vice of logic. Unity and extension are concomitant but not related, while one unit may be related to another unit and one extension may be related to another extension. Concomitancy and relativity must always be distinguished or there can be no sound psychology. The antithesis of this doctrine is sometimes held, which is an affirmation that the substrates of the universe are unknown reifications of number, space, motion, time and judgment. About unknown and unknowable things any assertion may be made, and all philosophies that are founded upon these reifications are therefore philosophies of disputation, as no two are alike. That which some great mind imposes upon his generation is by a succeeding generation gradually found to be more or less erroneous, and new philosophies are thus forever springing up, the one not founded upon the other; but gradually from generation to generation science establishes some things.

The relations which we are now to consider are those which are discovered when bodies are considered as particles. Quite a new class is discovered when we consider bodies as bodies.

As every particle of inanimate matter is a combination of four essential factors there are four classes of relations, namely: relations of plurality, relations of position, relations of path and relations of change, and these are all concomitant in number, space, motion and time. The same fact may be expressed in this manner. Relations of number are founded upon pluralities; relations of extension are founded upon position; relations of motion are founded upon trajectory; relations of time are founded upon change. Thus we have four classes of relations that must exist between particles. Then bodies have internal relations of particles and external relations when the body is considered as a particle in a higher body.

II

In a former chapter we spoke of the essentials of a particle of matter and considered them separately. Now we must consider them as they are related. There is a multeity of units, and plurality is founded upon units. The units are the terms that are related to constitute a plurality. A unit is unrelated or absolute in unity, that is, its unity does not depend upon others, but a plurality is dependent upon a number of related units; for example, the plurality may be ten; then ten as a plurality depends upon the units of which it is composed; nine is also a plurality, but it depends only upon nine units. A plurality is therefore a relation of units considered as a sum. Unity is constant only in ultimate particles. Bodies are combined, dissolved again and recombined, making variable units of plurality.

I am writing on a sheet of paper; it is one. With a match it is ignited and disappears; it is many. It was many before the conflagration, but many in one. After the combination these molecules though disembodied as a sheet of paper are still related to one another by all the concomitants, but now their more immediate relations are with the other particles of the molecules in which they are combined, while the new bodies thus formed have relations to one another of a higher degree or order in the corporeal world, for fixed internal relations constitute incorporation. Incorporation consists in the establishment or fixation of internal relations. When a body is disincorporated its particles dissolve their relation as one and assume relation with others to constitute new bodies or enlarge other bodies.

There is a great variety of relations between numbers. Numbers in nature are unified in orders of various kinds. The orders thus developed are multitudinous and quite beyond human comprehension. As the several units are compounds of individuals of lower units they are related to one another in infinite ways, as one is a multiple or sub-multiple of another. Thus we have one-fourth, one-half, equal to, twice, four times, etc. Mass is a sum of units measured in terms of force, and such units may become constituent parts in higher orders of units. One number is thus a measure of another. Out of these relations ratios and proportions arise. It seems unnecessary to enter into a discussion of the relations of numbers, as they are developed in the science of arithmetic and algebra.

III

Extension is exclusive occupancy of space. As there is more than one extension, and every one excludes all others, there is relative position. Thus we have positions derived from many extensions. Position is the relation of one extension to another. Space is founded on extension, for if a particle had no extension it could not be an element of space; a plurality of particles, each having extension, constitutes space. If they are in juxtaposition the space is the sum of their extensions. If they are separated by a medium, as for example an intervening fluid, the space is marked by their position and in this sense is related position; position, therefore, depends upon relation, but there can be no related positions if the extensions are annihilated. Extension is absolute, position is relative and space is absolute in extension and relative in position; extension is constant or persistent in ultimate particles.

In space one particle may be related to another in distance and in direction. These relations give rise to geometry and trigonometry and are the relations chiefly dealt with in astronomy.

In order that space may be discussed mathematically it must be reduced conventionally to number; this is done through the agency of measure. Then units of measure are devised giving rise to fractions and whole numbers, multiples, and sub-multiples, when it becomes amenable to the operations of mathematics.

IV

Speed exists in the unit of extension whether there be other units or not; speed, therefore, is unrelated or absolute. But the extended unit having motion must also have path, which is a change of position to others and variable by collision with others. It is thus relative. Speed is constant in the ultimate unit, which will be demonstrated in a subsequent chapter; but path is change of position in relation to others, and motion therefore is absolute in speed and relative in path.

There is persistence or indestructibility in the fundamental unit of extension and motion, but this unit changes its relation to other units in position and also in trajectory; the persistence is absolute and constant, the change relative and variable.

Motions are related to one another in direction and also in the positions of trajectories. Directions may differ in innumerable ways and paths may have innumerable deflections and thus trajectories may have innumerable variables. In order that direction and trajectory may be treated mathematically it becomes necessary to devise methods for the measurement of directions which are expressed in degrees and of lengths which are expressed in various measures. By these conventions motions are reduced to spaces and spaces to numbers, all giving an inconceivably great number of relations. But there are no motions without particles in motion, and there are no speeds without particles having speed, and there are no trajectories without particles having trajectories. There is no path without a particle having the essentials of a particle.

The science of the mathematics of motion deals with the speed of one and its trajectory, the speed of another and its trajectory, and of their collisions, and for this purpose it has to deal with the measure of their relations, and forever relation is considered and thus an illusion is sometimes produced, when motion itself seems to be wholly relation. Every particle of matter is in motion, and while this motion is absolute it is also relative. There can be nothing relative which is not also absolute, nor can there be anything absolute which is not also relative, and motion being thus absolute and relative it is quite proper to affirm this of motion, but it is not correct to affirm that motion is a relation any more than it is correct to affirm that motion is an absolute, if by these assertions it is implied that motion is one rather than the other; but if these assertions are made with regard to one correlative implying the other, then they are both correct. It is better form of speech to say that motion is absolute or relative when it is desired to call attention to one factor or the other, rather than to say that motion is an absolute or a relation.

The motion of particles is of such a nature that paths must impinge, and then collisions arise which give rise to impulse, or collision by which paths are deflected.

As bodies are incorporated in molecules of higher

and still higher orders, and through various molar forms as crystals, rocks, cells, phytons, plants, organs and animals and on into stars and systems of stars, each embodiment appropriates a part of the motion of its several particles or atoms. The molecules of the lowest orders have their motions, the molecules of the second order have their motions, the cell and the crystal have their motions, the earth has its motion and the stellar system has its motion.

The speed of every particle of matter is the sum of all the speeds of the bodies in which it is incorporated. Speed can never be increased or diminished in an ultimate particle; it may be increased or diminished in any one of its embodiments, but only by deflecting the motions in its other embodiments. This point is vital to a clear comprehension of the philosophy of science and is worthy of further illustration from the fact that it becomes necessary to rid ourselves of an illusion of sense. I see a bird perched upon a tree, then I see it flying through the air to perch upon another tree. The bird seems to have motion between the trees which it did not seem to have while perched on the one or the other; but the molecules of the bird before the flight had the motion of vitality, and in moving from tree to tree the trajectory of these multifarious minute motions are all deflected. The millions of millions of molecular motions had their trajectories changed. The bird itself was moving with the earth about its axis and with the earth about the sun, and with the sun about a point in Hercules. This is its astronomical motion. The change in the trajectory of the millions of millions of molecules was only the equivalent of the change in the trajectory of the astronomical motion of the bird. We know that all of these trajectories are changed; we do not know that the velocity or rate of speed of any particle of the bird's body was increased or diminished. If Newton's third law of motion, that action and reaction are equal, is true in the exact terms in which he stated it, then we must affirm that the speed of no particle was changed but only its trajectory. We do not see the astronomical motions of the bird nor its molecular motions. We do see the molar motions in flying from tree to tree and thus an illusion is produced that motion can be created or destroyed by the bird, and the persistence of motion seems to be a fallacy and the correlation of forces a fiction. We do not see the creation, continuance and annihilation of motion in the bird, but the deflection of astronomical and molecular motions as known by scientific investigation. This discussion is designed to show that motion is not a relation, but that one motion may be said to be related to another or related to any selected position.

V

Time is persistence and change, the persistence being absolute because it exists in the particle independent of other particles, and constant, for the particles cannot be annihilated. Change is relative, in that it inheres in the relations of the particles, and it is also variable, for particles are constantly changing their relations of position to each other by occupying a succession of positions. Thus time is absolute and relative, constant and variable.

The earth as a body changes the position of its particles by rotation upon its axis and thus passes through a series of daily events. It also as a particle changes its position in relation to the sun in a series of annual events. The position of the same body at one time may be related to the position of that body at another time; that is, its space relations may change.

As the motion of one body in its space element may become the measure of the motion of another body in its space elements, so the motion of one body in its time element may become the measure of another body in its time element. While particles are related to one another in number, space and motion, these relations are constantly changing so that they are also related in time; that is, particles are related to each other through their changes. A particle unmodified in its individuality may pass through a succession of changes by reason of its own proper motion determined by the motion of other particles. As the orbit of the moon around the earth may become the measure of the orbit of the earth around the sun, so the day may become the measure of the year. We have now found that numbers, spaces, motions and times are properties which can be measured, and through measurement which is conventional they can be investigated. We shall hereafter see how large a part of the scientific research pursued by man is occupied with these subjects. Quantity is the reciprocal of something else which is usually called quality, but in the course of this discussion it will be found that the term quality is badly chosen, that the real reciprocal of quantity is kind or class.

CHAPTER IV

KINDS OR PROPERTIES THAT ARE CLASSIFIED

I

Having considered the nature of the properties of a discrete particle of matter by reason of its own existence and the existence of others, we have now to consider how these relations are developed by incorporation. Still it is necessary only to draw upon the common stock of knowledge and deduce from it legitimate results which are easily understood. We have shown that the ultimate discrete particles are related to each other through pluralities, positions, paths and changes, and we have now to consider another method of association, for particles of matter are incorporated, and enter into fixed associations with one another by affinity, the nature of which has never been explained, although the association is well known. Every particle of matter under certain conditions seems to be able to choose its associate, and a group of such particles that have mutual affinities become compounded into that which is usually denominated a molecule. This association of particles in a molecule is not easily dissolved under ordinary conditions, yet if special conditions are provided the association is fickle and old combinations are dissolved that new combinations may be formed.

Then molecules enter into association with one another without cohesion in gases, with feeble cohesion in liquids, and a more tenacious cohesion in solids. A body thus considered of like molecules is called a substance.

II

In the combination of particles into molecules and other bodies, an interesting development of the properties is observed. Such a combination produces a new unit of a higher order. Here we find a new unity made such by combination, and it must be observed that it depends upon a plurality combined in one. It is therefore a new kind of unit. Thus a kind is developed by the combination of a plurality of units into one—a process familiar in the conventional units of arithmetic, where ten units of the lower order make one of the next higher order. That which is accomplished by convention in arithmetic, is accomplished by incorporation in nature. In this manner by combination the quantitative property of number in the particle becomes the classific property or kind in the molecule, and as there is a hierarchy of molecules and every one considered as a unit may become a particle in a higher order, we are compelled to consider it in this double and relative capacity, as one of many and as many in one. A molecule in its internal aspect appears as many; in its external aspect as one. Thus we have incorporated units, and these may be incorporated in a still higher order, and on indefinitely. There are many bodies of a kind and they constitute a class. Thus a class is a series or sum of a kind.

III

Ultimate particles, by reason of their extension and position, give rise to space; when they are incorpo-

KINDS OR PROPERTIES THAT ARE CLASSIFIED   33

rated positions are established in relation to one another, and thus a form is constituted. It is thus that space is developed into form by incorporation. It is seen that the particles of the molecule considered as such exhibit space with extension and position, while the molecule or other body also exhibits figure and structure. If we view the body from within as composed of particles, space is presented; if we view it from without as a body, form is presented.

Again, this same molecule may become a particle in a higher order of molecules, when it will exhibit space characteristics, and a higher molecule will exhibit form characteristics. Thus space is the reciprocal of form.

In my room there are desks, chairs, book-cases, books and many other articles. Their relations of position are relations of space. Were all these articles consolidated into one, so that one could not be moved without moving all, their relations would become relations of form. Contemplate a pile of cannon-balls; the relations of these balls to one another are relations of space; combine them into one body in such a manner that they will move together as one, their relations of space become relations of form also.

IV

It has already been asserted that a particle cannot lose speed. When we contemplate a molecule composed of particles in which relative positions are fixed, we are compelled to develop the thought one stage farther and conceive of them as still retaining their speeds. The concept of particles with relative positions fixed, and every one retaining its motion, can be realized by a consideration of facts presented by celestial bodies. The earth and moon revolve about a common axis which is within the periphery of the earth, and each retains its own speed while the relative position of each is preserved. But the sun and the earth revolve together about a common axis in the same manner, and the relative positions are preserved, while the relative position of the moon to the sun is indirect through the mediation of the earth. In like manner it can be shown that the relative positions of all the members of the solar system are preserved directly or mediately by a system of motion. Now, the solar system may be considered as the type of a molecule in which the particles retain their speeds and have their relative positions fixed by deflection in the motion of revolution. Yet the concept is not complete; for every one of the members of the solar system is rotating about its own axis. So that there is a complex system of motions within the solar system by which the orbs are kept within the theater of the system itself, even though the system as a unit may be revolving about some other point in the heavens; and the fixity of position of celestial particles is fixity of space relations about axes of revolution.

We do not know that the particles of a molecule move within the sphere of the molecule by a system of rotations and revolutions, though such a system can be conjectured; but whatever the system may be, it must accomplish the same results by confining a certain portion of the speeds of the particles within the theater of the molecule by a system of deflections, and, whatever may be the motion of the molecule itself in relation to other molecules, the speed of the particle must be partly taken up within the molecule, and it must then be divided between internal motion and external motion. Let us vivify this concept into greater distinctness.

Imagine a particle moving to and fro in vibration at the rate of millions of vibrations a second; the sphere of this motion is measured by the amplitude of the vibration. If the deflection is something less than one hundred and eighty degrees at both extremities of the vibration and on the same side, the particle will move off in the direction normal to the vibration. Its motion in the new direction increases inversely with the angle of deflection, until it reaches ninety degrees.

Now consider the speed of the particle in vibration when the deflection is one hundred and eighty degrees; then the total speed is represented in the vibration, but when the particle is moved in a direction normal to the vibration, the speed of the vibration is less by the amount of speed taken up in the new motion; thus the speed of the particle is divided between its two motions. In this manner we may conceive of the speed of the ultimate particle as being divided among the speeds of the bodies of a hierarchy in which the particle is incorporated. What we have shown about the speed of a particle in rectilineal motion is true of it in all forms of curvilineal motion. When one molecule collides with another each has its path deflected inversely proportional to its mass, for its mass is the sum of its particles, every one in motion and having a path of its own, and all of the particle paths must be deflected to a greater or less degree in order that the molecular path may be deflected. The force, therefore, with which one body deflects another and by which it resists deflection itself is the sum of the motion of its particles. Force, therefore, is a compound of motions. Thus motion in the particle becomes force in the molecule or other body. But the molecule itself may become a particle in a higher molecule, when its force becomes a motion which again must be composed.

V

We have now to consider the development of time by incorporation. It has been seen that time is persistence and change. The endless persistence of the particle is interrupted by changes in its relation to other particles, but when these relations are incorporated and become established as kinds, forms, and forces, time undergoes a development, for it then becomes causation as antecedent and consequent, or cause and effect. In ultimate particles collisions result in deflections and the changes which occur relate to paths; but the particles themselves are unmodified. When bodies are considered another set of relations are generated. With every collision the body may be modified, and a succession of these collisions may ultimately produce a great change. The change which bodies undergo in this manner is called causation. Thus, a body may be deformed or broken up, it may grow or decay when cause and effect are involved. Whatever happens to a molecule is distributed to its particles and is observed in its particles. If, now, we discover an effect and desire to learn its cause, we find the effect distributed to all of the particles which constitute the molecule and must go outside the molecule for its cause. This is what is known as the infinite regressus of causes.

The total cause of any event to a molar body stretches out through all the earth, and as the earth is a particle in the solar system the total cause embraces the sun and its planets and their satellites. Now, when we are considering an event as an effect, we are considering it as a change in the individual, but when we are considering the total cause of the change, we are considering the environment. The effect again becomes cause, which proceeds onward as a multiplication of causes distributed to all the environment. In the regressus of causes the total cause is multifarious; but we may from time to time consider any one of the effects of the total cause as the cause which may be varied in the production of an effect; then out of the effects of the total cause the one selected may be known as the special cause. This is the cause to which reference is made in common speech. An effect is observed in the explosion of gunpowder. We may consider the cause as the instability of the compound, the ignition of the powder with a match, or the purpose of the mischievous boy, etc. In like manner we may go on in an indefinite regressus to catalogue the causes of the explosion. When I am considering the conduct of the boy I attribute the cause to him; when I am considering the flame I attribute the cause to the flame; when I am considering the constitution of the powder I attribute it to the explosiveness of the substance. These are special causes as distinct from the total cause. Man comes to consider cause in this manner for a practical reason, for he interferes in causation for his own ends, and is forever searching for the most economic means of changing events

I am not familiar with any discussion of causation equal to that of John Stuart Mill in his work on Logic; but he failed to distinguish causation as an abstraction from force, form and kind. In his chapter on the Composition of Forces, he says:

> "I shall give the name of the Composition of Causes to the principle which is exemplified in all cases in which the joint effect of several causes is identical with the sum of their separate effects.
>
> "This principle, however, by no means prevails in all departments of the field of nature. The chemical combination of two substances produces, as is well known, a third substance with properties entirely different from those of either of the two substances separately, or of both of them taken together. Not a trace of the properties of hydrogen or of oxygen is observable in those of their compound, water."

In the chemical union of oxygen and hydrogen a new kind is produced as water. Here we have composition of kind; when causes are composed new conditions are developed; thus the oxygen and the hydrogen are found under new conditions of incorporation. In these new conditions there is a change in space relations, so that water occupies less space than the gases of which it is composed; thus the composition of kinds gives rise to the composition of conditions, but is not itself the composition of conditions as an abstraction. To discuss the composition of conditions it is necessary to discuss the very things to which Mill refers when he speaks of the development of new properties.

Heretofore we have used the terms total cause and special cause and have shown that the special cause is that one of a multiplicity of causes which is considered. Recurring to the illustration used before, it will be remembered that the cause of the explosion might be considered as the constitution of the dynamite, or it might be considered as the spark by which the powder was ignited, or it might be considered as the act of the incendiary, and in this manner we obtain the infinite regressus of causes. The considered cause may be any one near or remote in the infinite regressus. When any one is selected all the others become conditions; hence we have a cause and its conditions. So cause is related to effect and cause is also related to condition.

At this moment a man is climbing to the roof of my house. The cause of his climbing is a breach in the roof which he intends to repair. The cause of his climbing is his intention; the cause of his climbing is my request; the cause of his climbing is my knowledge of the breach in the roof; the cause of his climbing is the information given me by another that my roof leaks; the cause of his climbing is his desire to earn a fee; the cause of his climbing is his desire to purchase food; the cause of his climbing is his love of his family; the cause of his climbing is the hunger of his children. So we may go on forever to enumerate remote and distinct causes, and when we consider any one of them, the others become conditions which must be assumed as necessary to the operation of the selected cause.

We have considered teleologic causes; now we must consider genetic causes. The man falls from the roof. The cause of his falling is the misstep he makes; the cause of his falling is gravity; the cause of his falling is the greater distance of the roof than the surface of the earth to the center of the earth; the cause of his falling is the ascent to the roof; the cause of his falling is his coming to make repairs.

If any one of these conditions had been omitted he would not have fallen, and we can go on to multiply these conditions to an indefinite degree and discover that if any one was omitted this particular case of falling would not have occurred.

Why is one condition selected rather than another? This question might be answered by referring to the seriality of thought, which is the name for the law by which many things cannot be considered simultaneously. To comprehend all of the causes it is necessary to consider them separately; and while this is not a complete answer, it must be considered as an important condition to be understood that the answer itself may be understood. Man himself is a causator, and changes the currents of events in himself and in external nature. All human activities are designed to interfere with the course of natural events. Man bent upon the modification of events is forever intent upon the discovery of the most easily variable cause, and no small proportion of his energies are devoted to this discovery, and the invention of the way by which his discoveries may be made of avail. Hence it comes that particular causes are selected as those of most interest. Every act performed by man, every word spoken is an interference in the laws of causation and is designed as such. The artisan who repaired my roof interfered in the laws of causation by making the repair, but this interference can only be accomplished by the substitution of a new cause.

VI

We have now discovered that there is an additional property of the inanimate particle when it is incorporated, and that this is affinity. All we know of affinity is that it is the choice of one particle for another as its associate or is their mutual choice. Here we are introduced to the multitudinous phenomena of affinity, which can be explained only as choice. We must yet go on to consider other bodies than molecules to obtain a clearer idea of the nature of affinity itself.

VII

Class is the reciprocal of number. It is class in the body as kind and series, and it is number in the particle as unity and plurality. Form is the reciprocal of space, which is form in the body as figure and structure, and it is space in the position of the extensions of the particle. Force is the reciprocal of motion; it is force in the body as action and passion; it is motion in the particles as speed and path. Causation is the reciprocal of time; it is causation in the body as cause and effect; it is time in the body as persistence and change.

Number, space, motion and time are concomitant as they inhere in the same particle; kind, form, force and causation are concomitant because they inhere in the same body. These distinctions are radical, and must be firmly grasped if the argument herein presented is to be understood.

CHAPTER V

PROCESSES OR THE PROPERTIES OF GEONOMIC BODIES

The particles and bodies of the universe are fundamentally classified in six groups, as follows: (1) the particles of the ether, the science of which I call ethronomy; (2) the bodies and particles of the stars, the science of which is astronomy; (3) the bodies and particles of the earth, the science of which I call geonomy; (4) the bodies and particles of plants, the science of which I call phytonomy; (5) the bodies and particles of animals, the science of which I call zoönomy; (6) the bodies which are invented by men, the science of which I call demonomy.

I shall not write special chapters about ethronomy and astronomy, and shall consider demonomy in an incidental way and reserve it for a future volume; but I must devote a chapter severally to geonomy, phytonomy, and zoönomy, in order that we may discover something more about the nature of affinity and see if there are other properties which will require for their explanation more than the five essentials.

I

The earth is composed of four bodies surrounded by the ether.

First, there is a central nucleus constituting the principal mass.

Second, there is a crust of structurally disposed

rock surrounding the nucleus, the thickness of which is comparatively small.

Third, there is an aqueous body surrounding the rocky crust, through which the islands rise, the largest of which are called continents. On these islands there are many lakes and rivers which ramify into innumerable brooks, creeks and rills.

Fourth, there is an aërial mantle of air extending to a limit which is not well determined.

Fifth, these four bodies, one outside the other, in succession, are surrounded by the ether.

The earth is thus composed of encapsulated globes enclosing a nucleus and bathed in ether, to designate which certain definitive terms are needed. I shall, therefore, speak of the nucleus, the rocky crust or crust, the aqueous envelope or envelope, and the aërial mantle or mantle, and shall call them all spheres. For the sake of clearer distinction, these spheres may be called (1) the centrosphere; (2) the lithosphere; (3) the hydrosphere, and (4) the atmosphere. It must be observed that the ether is common to all of the celestial bodies, and perhaps penetrates them as it does the earth.

The centrosphere is the chief mass and has a density of 5.6. By reason of this great specific gravity, which is about twice that of the rocky crust, it is often supposed to be metallic. Geologic facts in a vast system lead to the induction that the centrosphere does not exist in the solid state; if it is metallic the weight reduces it to a trans-solid condition. To this condition the form of the earth testifies, as it is an oblate spheroid assuming the figure of a fluid under the combined action of gravity and rotation. These are facts which have led physicists to conclude that it must have a rigidity said to be equal to that of steel. This rigidity may be explained as a function of its rotation, revolution, and molecular motion, when the physicist and the geologist would be in substantial accord.

The theory of a metallic centrosphere seems adequately to account for the trans-solid state, as the metals are found to flow under pressure; but the molten material which from time to time is brought to the surface from the interior of the earth never reveals this metallic constitution. It may be that there is a zone of matter beneath the structural rock and overlying the metallic nucleus which is penetrated by heat, now here, now there, and only these molten rocks are extravasated; or it may be that the solid state is limited by heat in one direction and by pressure in the other in such manner that all rocks flow under great pressure as do the metals.

The stony crust has been revealed by direct penetration to a depth of more than six thousand feet, but it is indirectly revealed in many regions to a much greater depth, perhaps in extreme cases to fifty or sixty thousand feet.

The islands of dry land have all been beneath the sea at some time or other, and all show that they have been submerged more than once, some more frequently than others. During that portion of the history of the crust, which is the theater of geological investigation, these periods of submarine condition in one region always appear to be contemporaneous with periods of subaërial conditions in some other region. Thus there seem to have been regions of dry land and regions of ocean bottom coexisting with a large predominance of oceanic area.

The aqueous envelope covers the rocky crust over about three-fourths of its surface, and has an average depth of about twelve thousand feet, though in extreme cases the bottom of the sea is more than five miles below its surface, while in some few cases mountains rise to more than five miles above the level of the sea. It is certain that we are now able to study rocks which were deposited at depths much greater than that of the mean depth of the ocean, and there are many cases where rocks found on the summits of high mountains are known to have been deposited at great depths beneath the sea. Great regions of country are at one time submarine, and at another subaërial. These oscillations of upheaval and subsidence are oft-repeated in geological history, and the swing of oscillation seems to have been in some regions tens or scores of thousands of feet where they reach the maximum, and to be only tens or scores of feet at the minimum, so that the surface of the earth, in so far as it has been studied geologically, is found to give evidence of oscillations of level varying in these quantities.

These variations are geographically heterogeneous: one region may have its oscillation on a small scale, another on a large scale, the minor oscillations forming distinct geographical series and the major oscillations forming distinct geographical series; that is, one region has been subject during geological time only to minor oscillations, and another during the same time to major oscillations.

We must now more fully consider the nature of these movements. Sometimes upheaval is by anticlinal flexure, where the rocks are lifted along a line of upheaval and caused to dip away on either side in gentle or abrupt slopes which are sometimes beautifully curved; but such an upheaval often seems to be accompanied by a subsidence on the flanks. Symmetrical anticlinal flexures are not very common, but often one side slopes gently while the other is abruptly deflected. This abrupt slope is especially subject to rupture, in which case faults are substituted for flexures. Thus a block which dips gently in one direction has its margin, on the side of a fault, displaced as an abrupt escarpment. Blocks formed in this manner often careen upon their edges, so that the strata may become vertically disposed or quite overturned where the lower formed strata are found on top. Between careened blocks and flexed blocks no line of demarcation can be drawn: the same block in different parts of its course may be bent or broken, and the flexed blocks themselves be quite overturned. The rocks which are upheaved or depressed by faulting and flexing, one or both, are always found to be ruptured in line of the faults or flexures, and also transversely to them. This rupture is often minute, so that the sheets of rock are faulted and jointed and thus found in blocks of varying dimensions, but all very minute as compared with the widely spread formations from which they are broken. Thus the whole system of rocks, of igneous and aqueous origin alike, are broken into blocks by faults and ruptures, and still further divided by planes of deposition, so that the structural crust is a system of fragments sometimes with an area of many yards, other times with an area of fractions of inches. When we compare these blocks with the great area of the structural crust we find that it is but an accumulation of blocks that are to the formations what grains of sand are to the blocks. We must now realize that the structural crust nowhere has a continuous coherence; that faults, joints, and partings render it a vast body of minute and loosely accumulated fragments. All of this upheaval and subsidence with flexures, faults, joints, and partings seem to have been brought into this condition by intermittent convulsions often exhibited in earthquakes.

Having contemplated the lithosphere as a body moving in upheaval and subsidence, and shown what is about the maximum and minimum of these oscillations and their paroxysmal character, we are prepared to consider the structure of this crust.

In all geological ages volcanic eruptions have occurred and rocky material from the depths has been brought to the surface. Such appearances of lava at the surface have been very common in human history, and they appear to have been just as common in all the geological ages revealed by science. Lavas vary in chemical and mineralogical constitution, but this variation is within narrow limits. All of the mineral substances known to mankind appear, but are intimately mixed as minute ingredients. Lavas, therefore, are intimate mixtures of many substances, the average of which falls within narrow limits. It would appear from our present knowledge that the primordial surface of the earth was cooled lava and that lava has been erupted from time to time through all of the great geological ages.

Upon these cooled surfaces a new crust of rocks from below and rocks from above appears to have been spread. Wind waves and tidal waves are forever beating the lands and undermining the cliffs and distributing the materials beneath the sea. Then atmospheric agencies disintegrate the rocks and the rains wash the sands into the streams, which carry them into the lakes and into the sea. By many cognate processes the lands are worn down and the sea bottoms built up; the amount of detritus thus accumulated in zones about the meandering shores is great, so that in regions of maximum activity formations are accumulated thousands of feet in thickness.

The winds contribute to the material which falls into the sea; plant life also furnishes its quota; accumulations of vegetation are ultimately consolidated among the formations as beds of coal; and animal life adds to the marine formations, for corals, shells, and bones are all brought to be buried in the sand, and often extensive formations of calcareous matter are thus produced. From these sources the sedimentary rocks are brought to be mingled with the eruptive rocks and intercalated among them, while in turn they are thrust between the sedimentary rocks.

Layers of rock of sedimentary origin appearing as strata are commingled with other masses of rock of volcanic origin which come from the interior. Sometimes the lava flows under or between the sedimentary strata. When great masses of lava are found in these conditions they are called lacolites. Thinner sheets are called intrusive rocks. Beds poured over the surface are called coulées. The floods of lava come through fissures and fill them both below and above coulées, intrusions, and lacolites; such fissure formations are called dikes. Where the lava comes forth in volcanoes, the orifices are filled with molten rock which consolidates and are then called chimneys. Great bodies of lava are ejected by some volcanoes as scoria and ashes, and often the ashes are minutely comminuted; the expulsion of such material is doubtless due to the production of gases and vapors, especially of steam, and the comminution is probably due to the explosive actions of particles of water expanded into steam. Great volcanic cones are often formed by the piles of scoria and ashes which are extravasated, and the ashes themselves when highly comminuted are drifted by the wind, sometimes far away from the locus of eruption. Beds of ashes and scoria formed in this manner are called tuff. So the bodies of rock formed by eruption are commingled with the bodies formed by sedimentation, and all are known as formations. Both the sedimentaries and the eruptives undergo a further change, which to a greater or less extent obscures their origin, for the original formations are metamorphosed, that is, recrystallized and lithified; so that the planes of sedimentation are partly or largely obscured and the beds of lacolites, intrusive sheets, coulées, dikes, chimneys, and tuffs have a new structure imposed upon them, and are then known as metamorphic rocks.

An attempt has been made to define formations; now they must be considered in a new light.

The land areas have always been subject to degradation by rains, rivers, and waves, and the materials washed from the land have been carried into the sea and there deposited; thus the continuance of dry land area is comparatively ephemeral. Not only are the lands degraded in this manner, but when they reach the level of the sea they continue to subside; when above the sea they are speedily unloaded, but when brought to the level of the sea or nearly so the islands, though having their loads discharged, continue to sink. The regions which have received the detritus of the islands and are thus loaded by them, are elevated into the island or continental condition; thus land areas rise to be unloaded and then sink, while oceanic areas are loaded and then rise to become land areas. The extent of this upheaval and subsidence and the vertical movements, involved together with the vast transportation of material from land to sea, seems to be enormous when we contemplate the almost silent and unseen agencies by which it is accomplished.

In considering large areas of the surface of the earth, as, for example, the great continents or zones of archipelagoes, we reach certain generalizations of prime significance.

Regions of great denudation are also regions of great deposition, regions of great eruption, regions of great upheaval and subsidence, and also regions of great flexure and fracture; thus denudation and deposition, eruption and displacement (as subsidence and upheaval and as fracture and flexure) are correlated in this manner: that where there is more of one there is more of all; where there is less of one there is less of all.

Geologists have found no law, condition, or cause by which to explain these phenomena of the earth's crust as the law of gravity explains the constitution of celestial systems. The search for this law has been almost exclusively in one direction, under the hypothesis of a cooling and contracting earth, but with the lapse of time it has been found inadequate.

Attempts have been made to compute the amount of contraction supposed to result from the wrinkling of the crust of the earth in anticlines and synclines. It seems to entirely fail quantitatively. Contraction does not seem to be an explanation of all or even the chief phenomena which we have briefly set forth. When this hypothesis was considered, flexion seemed to be the chief method of displacement; now we know that fracturing and faulting is the chief method in regions of maximum action. When inclined rocks are studied they seem to have been stretched, as evidenced in the elongation of particles transverse to the strike, and they seem further to have been stretched by the opening of fissures and joints. Altogether it may be affirmed that displacement does not teach the doctrine of a contracting earth, or, if that statement is too strong, it does not give evidence of a sufficient contraction necessary to the hypothesis, and it also fails to explain the concomitant phenomena.

With this hypothesis another is associated, namely, that the centrosphere of the earth is metallic, for which no vestige of inductive evidence has yet appeared; and the stupendous fact remains that the centrosphere has more than twice the density of the crust. All eruptive rocks which come into the purview of science are found to have an average constitution which is about the same as that of the sedimentary rocks. It is found by experiment in the industrial arts that under pressure metallic and other substances flow; and geology teaches that all of the other rocks are secularly deformed under differential pressures, so that rocks highly metamorphosed in this manner are twisted, contorted, and kneaded into new shapes. Finally, there is now abundant geologic evidence to show that the faulting near the surface appears as flexure at greater depths, and finally that flexure appears as molecular readjustment at still greater depths, expressed in slaty structure where the particles of the rocks are rearranged in parallel planes.

The metals of the normal condition have great density, but in a pure condition are found only in exceedingly minute quantities; all the other rocks have a small density. If we now assume that all rocks flow under pressure, that the critical point is variable and that the modulus of compression is also variable, being greater for the lighter rocks and less for the heavier, and that this modulus is greatly accelerated at the critical point, we have a law which will regiment the facts of geonomy as the facts of astronomy are marshaled by the law of gravity.

Under this theoretic law of the accelerated modulus of compression at the critical point for different substances, subsidence and upheaval are explained. The reassumption of constitutional structure in crystallization and glassy lithification necessitates expansion, and thus upheaval is explained. When lands rise and are denuded, the process of relithification in the centrosphere continues upheaval and exposes the lands to further upheaval, and this process goes on until an equilibrium is reached at the epoch when the land is brought to the level of the sea by degradation. On the other hand, as land is loaded the subjacent crust rocks are brought within the zone of accelerated compression, and this process continues while the loading continues until it is brought to a close at the epoch when the land area from which the detritus is taken is brought to the level of the sea and transportation ended so that loading ceases.

Universal contraction by cooling must still be postulated as an agency for the destruction of equilibrium, or perhaps we may find this agency in astronomical conditions; but some such agency is necessary for the continuation of the process. But the changing of material from the interior to the surface and the changing of load from one district to another by transportation under the law of the accelerated modulus of compression is the principal agency of upheaval and subsidence.

This doctrine was proposed several years ago by myself, but has received little attention except among a few geologists engaged in this branch of research; from its reception by these gentlemen I am encouraged to repropound it.

The hydrosphere requires a little further consideration. The water evaporates from the surface aided by a variety of conditions which cannot here be considered; as vapor it floats in the air; then the rocks by atmospheric agencies are reduced to dust and blown by the winds and seized by the vapor, so that particles often become the nuclei of raindrops. The falling of the water restores the particles of dust to the crust. On the other hand the water penetrates the rocky crust by the innumerable fissures which have already been described and along the partings of the rocks and among the sands of which they are composed. In a condition of vapor it is probable that it penetrates through all of the stony crust. Thus it falls into the earth by streams, by capillary channels, and into the metamorphic masses at great depths, where it assumes the rôle of an agent of rearrangement in crystallization. There is much evidence to show that this finally becomes the agent of explosion when the rocky masses are thrust by the weight of superincumbent rock into the centrosphere, for this seems to be the explanation of the tufaceous material thrown out by volcanoes. This penetrating water becomes the agent of another process which goes on in the crust on a vast scale, for the waters, especially when they become thermal, dissolve certain substances and redeposit them as they are evaporated above and as they become waters of crystallization below. Especially are the metals treated in this manner, giving rise to metallic lodes by solution in the water and their subsequent evaporation and crystallization. The formation of mineral lodes in this manner is a long but interesting chapter in the story of geology.

We now have a condensed but perhaps sufficient account of the structure of the earth in spheres and their interaction in the production of formations. We must now consider these formations abstractly in the light of the essentials as they are changed in relations of quantities and categories into formations.

II

In the deeply seated rocks substances are transmuted by recomposition, secularly accomplished by changes in heat and changes in pressure which produce chemical reactions. As the rocks sink under the materials piled upon them by extravasation and deposition, they are faulted and jointed, and this permits the water to flow in underground courses; these flowing waters dissolve certain substances on their way down, and deposit them again, filling the joints and fault seams with deposits accumulated from higher grounds. As the upper and lower surface of the crust is approached the rate of change in the substances is increased until these surfaces are reached. At the upper surface the disintegrated rocks form an overplacement of soils which undergo such chemical reaction that the substances of vegetal life are produced. This material, exposed for longer or shorter periods, is transported by streams to lakes or to the sea and sinks to the bottom, where it is recombined into various substances, especially as carbonate of lime, chloride of sodium, other salts, clay and coal. All of this transmutation is a numerical change in the relation of the atoms to the molecules of the substances developed. Let us call it metalogisis.

The new substances which appear in the changes wrought by the agencies which have been described are segregated in the deeply seated rocks as crystals. Those which are formed in the fissures appear as bodies of ore and those that are washed from the surface and deposited at the bottoms of lakes and seas are arranged in strata, but as the waters themselves dissolve the substances of the surface they are often recombined and crystallized. Thus it is that the new substances are segregated and the new mass of comminuted material has the new kinds developed in this manner, separated more or less distinctly from the kinds of the original mass. Thus metalogisis is the genesis of new kinds and their segregation by a succession of changes.

Thus we see that in the processes that go on in the crust of the earth new kinds of substances are developed and new kinds of formations produced, and the chemist finds these substances to be arranged in series, and the geologist finds that sedimentary formations are arranged in series. So in geonomy, kinds are developed into series.

III

With the change in kind comes the change in form which is accomplished by minute increments. When the mineral substances are recombined in the deeply seated rocks they are slowly metamorphosed by recrystallization and rearrangement in slaty structure. The ores are deposited in mineral lodes and to some extent crystallized. The sedimentary formations are arranged in layers or strata, and are thus seriated. Heavier and larger materials are sooner deposited, lighter and smaller materials are slowly thrown down, and the currents of the water carry them farther away from the shore; thus there is an assorting process which is still farther extended by the deposition of materials in solution. In this manner the structure of the rocky cellate is constantly undergoing metamorphosis.

The slates are seriated and the sedimentary strata are seriated. Thus kinds are seriated as forms revealed in structure and figure. The elements of structure are set forth in a more elaborate form in structural geology when slaty structure, lode structure and stratified structure are the themes, and where flexures, faults, fractures and displacements are set forth in describing the structure of mountains, plateaus, hills and plains as slates, lodes and strata, giving figure to the topographic features and the endless variety and beauty of the topographic landscape. This figure is revealed in valleys with stream channels and canyons.

In plains that are sometimes baselevels being asymptotic and sometimes surmounted with monadnock elevations, in plateaus with abrupt escarpments and fringing hills, in mountains which are often systems of ridges carved by gorges into peaks or elevated as volcanic cones, all spread with a parterre of forest, meadow, field and flower through which flow rivers, creeks, brooks and rills, where cataracts and cascades are found and where fountains issue from the rocks and lakes are nestled that mirror the vegetal-clad shores; while away to the polar region the ice gathers, and the glaciers break into icebergs and float down the sea, or following the land, carve valleys and build moraines. All these things and many more constitute the theme of physiography, which is a description of the figure of the oblate spheroid. A succession of changes of form we call metamorphosis.

IV

In the change which comes in the development of the rocky cellate, forces become energies; that is, pari passu with metalogisis and metamorphosis there is metaphysisis; and metaphysisis is energy and work as reciprocals. The same fact is sometimes expressed in another form. The spherical members of the earth and the formations of which the crust is composed exhibit strains and stresses in their interaction and these strains and stresses produce changes. The varying heat of the ether by contraction and expansion rends the rocks and is an agency for their disintegration. The ether evaporates the water, the wind carries it about and fills the air with dust, and the dust and vapor again fall to the earth as rain, and the falling becomes a process of disintegration in part, but mainly an agency for the transportation of material of the rocky cellate by sheets of water into streams and by streams into the larger bodies. Then gravity acts as a process, throwing the load of transportation to the bottom in assorted layers. Then the percolating waters exhibit new processes of transmutation. With all of this there go the processes of strains and stresses in the rocks themselves, some formations being relieved of pressure and others having pressure added, and all these work their changes. Then there are the processes of extravasation consequent upon the relief and addition of the strains and stresses.

All of the processes which are here but partly enumerated are intermittent. The ethereal processes change hourly and daily with the longitude, and vary with the latitude. The winds blow and are calm; evaporation goes on until critical conditions are reached when storms fall; floods are also variable, and floods produce effects in geometrical ratio. The pressures of formations have their accelerations intermittent, so that stresses are revealed by earthquakes, and the fractures caused by earthquakes produce the channels for eruption, and add to pressures and stresses.

There is a change in hydrostatic pressure of such importance that it must not be neglected. The waters that are wedged between the stony blocks and thrust into pervious strata and absorbed into all of the rocks by processes of crystallization are subject to the same intermittent activities.

Again, on the streams of great floods, great blocks are loaded, and as these blocks become larger they are the more efficient as hammers in the corrasion of stream channels both vertically and laterally; so that glaciers load themselves with rocks and become the agencies of corrasion by which valleys are carved.

All of these processes are the work of gravity, heat, light, electricity and magnetism, and combined they produce a set of chemical changes which, as a mode of motion, we call chemism, which must be distinguished from affinity, for affinity means choice, while chemism means energy, and valency expresses numerical proportions. Heat produces expansion, gravity produces contraction in the materials of the rocky crust, and, conjoined, they produce chemism. This geochemism is the fundamental energy.

Stresses and strains are produced in celestial bodies as exhibited in their spheroidal structure, but chemism appears in the particles of which celestial bodies are composed, and at present we cannot study these particles in any other celestial body than that of the earth; chemism is a new mode of motion exhibited to us only in the earth, though we may conjecture that it exists in other globes if we could examine into their geonomy. A succession of changes of force is metaphysisis.

V

We have next to consider a succession of causes and a succession of effects. The rill rolls down the declivity; by the process of corrasion a channel is cut, and this effect is a continuous deepening of the channel. The cause is a process and the effect is a process; a serial causation, therefore, is a double process, one of cause and the other effect. The water on its way down the rill transports the abraded rocks; thus there is a constant process of cause in the flowing of the water, and a constant process of effect in the transportation of the load. When the rill reaches the foot of the declivity by the change of grade in the stream it is no longer able to carry the load, and it is deposited. The constant process of discharge from the water results in a constant process of deposition upon the bottom. It is in this manner that causation is continuous, and such a causation is a double process. A serial force is a process. In energy the work done by force is proportional to the time in which the force acts, but in the process this law does not necessarily obtain, for cause is not wholly a question of force but it is also a question of form and kind.

The rate at which the stream corrades its channel is due in part to the mass of the water and the declivity of the stream, that is, energy, but it is also dependent upon the form of the rocks and their chemical constitution. If they are easily disintegrated they are loaded the more, and the sedimentary particles as the instruments of corrasion are multiplied. Much depends upon the constitution of the rocks. If they dissolve in minute particles they corrade less; if the particles are larger they corrade more. Thus the rate of corrasion is a function of force, of form and of kind, and hence there can be no equality between the work done as an effect and the energy as a cause. Again in transportation of the material the rate of transportation depends upon the rate at which the supply is furnished, and not upon the force of the waters, for the supply is load and the load adds its own weight to the gravitating energy. The condition of fineness in the particles constituting the load will greatly aid transportation; the larger particles will sink sooner, the smaller particles will be carried farther; the deposition will in one place be of large particles, another place of small particles; hence a new effect is produced, that of sorting the material. It has already been shown that causes are multifarious and run into an infinite regressus, and between no one of these causes and the effect does there exist the relation of equality, and because the causes are disparate from the effect there can be no equality between the cause and effect.

This is one of the strange fallacies often met, and its origin lurks in the term action and reaction when bodies in motion collide. A and B are two bodies in motion; they impinge and are mutually deflected. Now if we consider A before the deflection and after it, we have the two directions separated by an event. The same is true of B before the collision and B after it. At the collision there is a double cause involved in the incident motions of A and B before the collision, and a double effect in their reflected motions. As force there is a mutual action and reaction, then there is equality existing between them. As cause and effect there is a mutual causation. The angle of incidence equals the angle of deflection; that is, there is equality between angle and angle as relations of form; but this is not a relation of cause and effect as such. We must find the cause of the collision, and then we may find what the collision causes. Change the conditions in the two particles; let one of them be easily crushed and the other not, then one ball will rebound and the other will be shattered. Now action and reaction as force will still be equal, but cause and effect will be different conditions; one body has its course changed, the other body is shattered into fragments, and these fragments take different courses. Thus it is seen that between cause and effect equality cannot be asserted. There is no equality between a word of command and prompt obedience to the command. There is no equality between sunrise and the opening of the morning-glory; there is no equality between the story of the Bonnie Brier Bush and my emotion. It is always abuse of logic to assert that equality exists between cause and effect, although the first mode of causation has that characteristic as change of direction.

When we consider force as force there is always equality between action and reaction; but when we consider force as cause, then no relation of equality exists between it and effect. A unit of force may raise a hundred pounds to a given height; two units of force may raise two hundred pounds to the same height. Thus the work is proportional to the force; but we are not considering a relation between forces but a relation between cause and effect. When in lifting the weight we consider it as an effect and wish to refer to its causes, they are found to be in the machinery by which the effect was produced, in the application of the force to produce the effect, and in the origin of this force. That is to say, whenever we are examining the relation between cause and effect we are examining into conditions or states and not into equalities or inequalities of force. When a weight of a hundred pounds is raised a unit of altitude, the effect is a new position, and the force employed, which was one of the causes of the new position, is an action equal to the lifting of the weight as reaction. But the cause might have produced a very different effect than that of lifting the weight; the effect might have been the breaking of the rope; then the cause is the force and the effect the fracture. Cause and effect are not relations of force to force, form to form, nor kind to kind, but they are relations of time to time as they are affected by force, form and kind. There can be no cause without force, form and kind; that is, we cannot analyze cause but can only abstract it. We cannot put cause in one basket, force in a second, form in a third, and kind in a fourth; and this is only a repetition of what I have said about unity, extension, speed and persistence.

A process of causality is here called metagenesis and a series of changes are produced.

We have now seen that the four essentials are still represented in the processes of geonomic bodies, and we also see the action of affinity in these bodies, and affinity itself is never revealed except as choice.

CHAPTER VI

GENERATIONS OR PROPERTIES OF PLANTS

I

We are yet to follow properties through higher degrees of relativity. For this purpose it becomes necessary to examine the relations exhibited by plants in metabolism, growth, vitality, and heredity. Plants are not wholly disparate bodies, but rise by a discrete step or degree in relativity not exhibited in ethronomy, astronomy and geonomy. So that not only are the properties in those realms found in this new realm, but in addition a new set of relations which we denominate generations. We have, therefore, to examine those characteristics by which plants are distinguished from geonomic bodies, of course in only a general and summary manner.

In plants new kinds appear by chemical recomposition. A new substance, protoplasm, is constituted, being organized of many molecules of different kinds, which again combine with other substances. These molecules seem to be still further arranged in different proportions, by which the new plant substances become many; the formation of these substances is called assimilation.

The many substances of plant tissue have a secular development which is growth in size and form. The period of existence of the plant body is limited, and at death returns to simpler conditions; this return is decay, and belongs to the grade of processes. In growth the plant undergoes a change of increasing relativity, and in decay returns to a simpler state of relativity.

During growth, which is an increase of form and structure by a succession of changes, it also exhibits a new mode of motion, which is vitality or life, and the cessation of this activity is death, when the plant returns to the geonomic world by decay. But assimilation, growth and life are continued from one generation to another, and imply time from period to period. This time is occupied in making changes, and causation is metagenesis. Now a new element of time appears, for by producing germs and thus multiplying individuals like itself the same stages of metabolism, growth and life observed in the parents are repeated in the offspring. This new element is heredity, in which the offspring inherits the potentiality of the parent as it is restricted within certain narrow limits to careers of metabolism, growth and vitality similar to that of the parent.

Thus generations are generations of processes. The processes are assimilation, construction and destruction, growth of form and structure, vitality exhibited in endosmosis and exosmosis, and finally processes are repeated by heredity represented by parents and children.

In this grade of concomitants it must be observed that there can be no assimilation without growth, no growth without vitality and no vitality without heredity.

Indeed, as we go on to contemplate the concomitants that appear by increasing relations, it becomes more and more evident that one cannot exist without the others, and that abstraction must always be distinguished from analysis. It becomes possible to treat the whole process of plant formation as assimilation or as growth or as life or as heredity, and yet we distinguish these concomitants in thought. If we treat of the assimilation of plants, the phytology of plants, the vitality of plants or the heredity of plants through germs, we seem to take the whole subject in view, for the concomitants are not disparate but only abstract in consideration.

II

I define constructive assimilation as the building up of protoplasm, a compound composed of many molecules, and I define differentiating assimilation as the recombination of protoplasm into other substances which are simpler compounds. These simpler substances are composed not only of some of the molecules of protoplasm itself, but also of other substances, and are used for various purposes in the economy and structure of the plant. In these recombinations a surplus of substance is found which is excreted by the plant in two ways: first, as water which is imbibed and used as a vehicle for other substances, for the amount of water is in excess of the amount ultimately used in the tissue of the plant, and is excreted by transpiration; and second, as carbon-dioxide, for the oxygen of the air unites with an excess of carbon, and it is then excreted by respiration. Thus protoplasm is the basis of the tissues of the plant; but to make these tissues it must be recombined into different substances which are newer compounds, and new substances not found in protoplasm are necessary therefor. The water which is necessary for protoplasm is furnished together with an additional amount which becomes the vehicle for the new substances, and the surplus is excreted. In the building of new substances oxygen from the air is needed to dispose of some of the carbon, and this office is accomplished by respiration. Imbibition of water by the roots furnishes the material for assimilation both constructive and differentiating, while respiration in the leaves furnishes the oxygen necessary for certain chemical changes. We must now consider the substances produced by assimilation.

The plant is a chemical laboratory of exceeding complexity, where all of the operations are carried on with marvelous deftness and delicacy, and with a system of chemical paraphernalia adapted to the operations of microscopic life. The entire plant is engaged in these operations as long as life lasts, sleeping in partial rest by night and hibernating in semi-torpidity during the winter, but carrying on its operations in full vigor when the sun is genial. Assimilation deals with particles so minute that even the eye of the microscope cannot see them, and they can be known only when aggregated in masses as material for use or as products, but the operations are carried on particle by particle in such a manner that what is and what becomes reveal the method of becoming only to the eye of reason; thus ultimately all chemical knowledge is the product of inference. Nevertheless this inferred knowledge is erected upon a foundation of consciousness as revealed by the senses, and the ultimate proof of the validity of the inferences is the multiplication of facts as they are accumulated in vast numbers by history and attested by the verification of prophecy; finally, as the facts are resolved into laws their congruity is made evident. The love of truth born of the generations of thinking minds forever engaged with the materials of consciousness in the process of inference, ultimately establishes a habit and love of truth that submits every judgment to the tribunal of congruity, the court of equity which every man erects in his own soul. This is the supreme court of judgment.

History may decide and prophecy may confirm, but these decisions are annulled if the court of congruity finds them contradictory. Experience in the laboratory may pile up facts, prophecy in the laboratory may be fulfilled in multitudinous cases; but under the decrees of the court of congruity if any incongruity appears the chemist is turned again to his experiments, resting assured that somewhere his facts or theories are wrong, and he plunges into his labors to reach peace only when congruity is found. To a man who has not devoted his life to chemical research and has familiarized himself only to a limited extent with the history and theories of chemistry, the vast body of experiments, the innumerable verifications of prophecies and the congeries of congruities which have developed since Dalton propounded the atomic theory are such a monument of accomplishment by inference and verification that they appear as a pyramid of truth.

The fact to which especial attention is called is this: That the laboratory reveals in the substances of plants innumerable new kinds and these new kinds are found in series.

III

The plant is a laboratory for the evolution of many substances; but as the particles of which they are composed have number and as they are arranged in numerical, that is, molecular orders, they have at the same time extension, and their arrangement implies that they are placed in forms. Thus having considered generations of kinds, we are led to the consideration of generations of forms, and the forms which we have to consider are forms of cells, forms of tissues, forms of phytons and forms of plants. Plants also exhibit forms of crystals; the crystallization is fundamentally a theme of geonomy, so on the very start of this subject we are confronted with the fact that the plant exhibits the concomitants of lower relativity, but for present purposes we may neglect them.

The normal and developed cell has three concentric envelopes which may be called blasts, the whole enclosing a nucleus, so that the structure which we found in the earth as spheres is repeated here as blasts. These are the exoblast, mesoblast, and endoblast. Some plants are single cells, other plants are aggregates of loosely attached cells joined together as threads or as webs of threads as in the slimes, but in plants of a little higher grade these webs are consolidated by a woof of plant tissue as in some of the lichens and seaweeds.

The tissues are consolidated and modified cells. Then tissues are differentiated, exhibiting different structures; different structural tissues are again related and modified for the performance of functions as phytons and the phytons are systematized to constitute the plant, but the phytons are differentiated for special functions and we have the roots for imbibition, the leaves for respiration and transpiration, the circulatory apparatus for transportation, the floral phytons for reproduction and the protecting apparatus for the external covering of plants.

A system of phytons constitutes the higher plants. In the history of plant life the morphology of plant phytons is an important part of the science of botany, for the forms of phytons undergo a succession of changes, the investigation of which vies in importance with that of the chemical development of kinds to which we have heretofore alluded. When the different classes of plants are examined in this respect, the succession appears in the development of classes, those plants of the lower classes passing through morphologic stages which are repeated in higher classes and continued to still higher stages, so that the plants of the highest class practically include all of the stages in succession as exhibited in the order of the lower classes. While some research has been devoted to this subject, much more requires to be done.

IV

That which we call chemism is one of the concomitants of process and is here transmuted into vitality. Vitality is chemism internally controlled by the plant in obedience to the laws of heredity, and externally controlled by heat, gravity, and strain which produces stresses. Thus vitality is a new mode of motion. We must here remember that motion as speed is inherent and constant in the particle and that motion as path is always determined from without, but the particles within the body are all external to one another, and therefore the direction of motion which is internal to the body is in obedience to the laws of heredity, and the direction of motion which comes from without the body is heat, gravity and strain. Heat and its modification as light and perhaps as electricity and magnetism play an important rôle in vitality, which has been subject to much investigation by the observation of nature and artificial experimentation. The vitality of the plant is accelerated by heat, and becomes torpid when it is insufficient. Certain chemical processes, like that of the production of chlorophyll, are dependent upon light. Doubtless gravity exerts a direct influence upon the functions of the plant, but this influence has had inadequate examination. Stress and strain are exhibited as endosmosis and exosmosis, exhibited to us in the circulation of fluids through the membranes of the cells, and is an important theme in the physiology of plants.

V

As the plant germinates the motions of its particles in change are directed by the preëxisting constitution of the germs; assimilation, therefore, is a directed motion, and as changes in assimilation and growth proceed the continued motion of vitality is controlled by antecedent conditions. In this manner the plant must pass through the same phases of assimilation and growth through which the parent proceeded; thus conditions are imposed which constitute causation; but there are other causes than those inherited, for the germ may not grow at all; it may not get footing in the soil, it may not find sufficient moisture, or the moisture may not contain other necessary ingredients. When it starts the frost may nip it, the sunlight may fail it because of an overhanging shade, herbivorous animals may devour it, man may dig it up. All of a multitude of conditions are necessary that a plant may mature; and these causes may be traced to the ultimate supply of food, as effecting the assimilation, to external forms which cast destructive shadows or protect from destruction, or they may be traced to external forces, so that there are heredity conditions and environmental conditions.

The plant is thus subject to inexorable conditions by its inheritance, and these conditions restrict its growth to the course pursued by its ancestors; but heredity is not the only factor of causation involved; the environmental factors may succeed in preventing, arresting or modifying the development of the plant. When the plant arrives at maturity and produces other germs, they also are subject to the laws of heredity; but the inheritance which they receive has accumulated in the development of the parent. Thus as generations pass there is secular development.

VI

Metabolism implies affinity, and again we have the problem of its nature in plants. It has often been surmised, and sometimes taught, to be choice. It seems to be the same thing in plant life, but there are other phenomena which appear in plants which suggest that the ultimate particles have not only the power of choosing their atomic and molecular associates, but they also seem to have the power to a limited degree to choose the attitude of their phytons toward external objects in space. Thus certain phytons seek the soil where they may perform the function of roots, and others seek the air where they may perform various functions in the plant life. These subaërial phytons seem to be able to direct their course toward different objects, when they require support, as in the case of climbing plants, and the leaves seem to be able to open or close in order to adjust themselves to conditions of light and darkness. The investigations into these functions of the plant are numerous and interesting, but they have been pursued mainly with the purpose to account for them as of a mechanical nature. Yet the problem remains: Have the plant elements the property of choice? If they have such a property they must also have consciousness.

We find in plants the same essentials: unity, extension, speed, and persistence as they are compounded into the properties of number, space, motion, and time, and as they are further developed as time, form, force, and causation; we also find the fifth property of affinity, which now seems to be choice even more plainly than we have found it in other bodies.

CHAPTER VII

PRINCIPLES OR PROPERTIES OF ANIMALS

We are yet to consider a higher degree of relativity than that exhibited in the bodies which we have heretofore examined. This higher degree is the discrete degree observed in animals. Plants have assimilation, which is both constructive and differentiating. In animals this rises to a high degree of relativity in that assimilation, both constructive and differentiating, is coincidently accompanied by destruction of the part that is reconstructed. The plant assimilates until its growth is complete, except in the higher plants in which the leaves drop from time to time and are returned to the inorganic world, and except in the same higher plants germs are given off which may be returned to the inorganic world, or continue as new plants when new plants are developed, but the trunk of the plant remains while it grows, and is returned to the inorganic world only when it dies. The animal assimilates and coincidently with this assimilation gives up a part of its material to the inorganic world. This is what I call metabolism, which is both constructive and differentiating of the material wrought into the structure of the body, while at the same time a part of the material of this structure is disintegrated and returned to the inorganic world. Thus the animal dies in part that it may live as an individual, and if it ceases to die in part it ceases to live, and when it ceases to live through death, it dies altogether and returns to the inorganic world. In other words we may say that in the plant phytons are dropped and renewed, but in the higher animals, organs which are homologous to phytons are not dropped and renewed, with minor exceptions, although molecules of the organ are discarded and coincidently new molecules take their place. By metabolism, therefore, we mean something higher than assimilation by a discrete degree of relativity.

So the animal grows not only by molecular additions to its substance, through which its size is increased, and whereby structural material is added, but the structure of the animal itself is constantly undergoing a change. Throughout the whole animal body a reconstruction is forever in progress; and this continues even after growth ceases as long as life lasts. This is the new principle of form, which is reconstruction.

Every particle of matter has speed, which cannot be increased or diminished, but the particles of inanimate matter seem mutually to direct one another's paths except in the case of incorporation, when they seem to be directed by affinity, the nature of which is not fully explained. The animal has a new power by which it determines its own path as a body; thus it can direct its own course. The animal is encompassed by an environment out of which it cannot pass but within which it can move as it chooses. With some animals this environment is the atmosphere, with others it is the hydrosphere, while other animals are fixed to the rocky sphere and have their movements greatly restricted in the hydrosphere or the atmosphere. Of those animals that have three degrees of freedom in the two outer spheres many are restricted by climatic conditions. The mode of motion by which animate bodies are capable of this higher degree of motion I call motility, which is self-directed molar motion or self-activity.

As this self-directed molar motion appears in the animal it enlarges its theater of action, being able to seek a new theater in which its self-activity may be employed. Thus the animal, no longer confined by a narrow environment, is able to invade a new region and exercise itself there. The animal can go from one environment to another in search of new conditions, changing the environment by its activities and taking advantage of the new environment by receiving the effect which the new environment produces. It is necessary for the plant to remain in a fixed environment and to act only when it is acted upon, but the animal may seek an environment more congenial and conducive to its wants, or ideals of good; thus it may escape evil on the one hand or acquire good on the other. It may choose its activities. This I call self-activity, which is force of a higher degree of relativity than that observed in plants by a discrete degree.

The animal, like the plant, has heredity, and its ancestors are the causes of its activity from which it cannot wholly escape. Its self-activity is therefore only within the compass of its hereditary activity; but while it has hereditary activity it is also subject to environmental actions, which are also causes from which it cannot escape. But as it chooses its environment within degrees of freedom, the environment is not wholly inexorable. Thus if food does not come to the animal the animal may go to the food. When the storm comes it may escape its action by seeking shelter, and in multitudinous ways it may choose the activities in which it was engaged, and choose the actions of others to which it will submit.

The lower forms of plants multiply by subdivision, but in the higher plants they multiply by sexual conjugation, the different sex organs being produced in the same plant or in different plants. In these higher plants the conjugation is adventitious in that the pollen must be carried by the wind or by insects or other agencies from the male to the female plant. But the higher animals have the power of choosing their mates, so that the continuance in generations is controlled by volition.

In plants, male and female germs, as particles, conjugate or choose one another. In the higher animals male and female bodies conjugate as bodies. This conjugation is accomplished by the mutual choice of the individuals as bodies, and the mutual choice of the individuals as bodies involves the consciousness of both, and this consciousness must have expression, and this expression is language. Hence reproduction in animals is dependent upon the mutual choice of animals, which choice is expressed in language in some form or other. Here we have a discrete advance in degree of relativity in reproduction which we call expression.

In animals we clearly find a fifth property which we cannot ignore, and which ultimately we shall find to be strangely like affinity. This property of the animate body permits it to form judgments about the nature of environments, and then it may form judgments about the good and evil of these environments in relation to itself. Judgments grow into concepts as judgment is added to judgment by experience. Thus a body of judgments is formed concerning every object in the environment which grows by increments of judgments. These concepts, which are the creation of the animal, constitute the fifth principle which we have to consider. We have therefore metabolism, reconstruction, motility, expression and conception with which to deal in the consideration of animal bodies.

These five principles exist in the lowest protozoa or unicellular animals. The evolution of animal life is the development of organs of metabolism, reconstruction, motility, reproduction and conception.

The five systems of organs are concomitant in the same animal body. They are also concomitant in every organ of the body, so that when we describe organs it becomes necessary to consider their concomitants. An organ may have the function of one concomitant, but it has the essentials of all the concomitants, for they cannot be dissociated, as we have many times seen. A certain part of the matter of the body is set apart to perform a specialized office for the other parts; and this specialization is accomplished by assigning a function to the essentials or concomitants severally. It is thus that there are five systems of organs; the first for metabolism, the second for reconstruction, the third for motility, the fourth for reproduction, and the fifth for conception. Thus we have the digestive apparatus, the circulatory apparatus, the motor apparatus, the generative apparatus and the conceiving or thinking apparatus. These apparatuses are completely concomitant with one another; so that every organ of the body, whatever function it may peform, must also perform the other four functions in an ancillary manner. When we are considering an organ we are compelled to consider a dominant function with four ancillary functions; or it may be stated in another way: an organ cannot act but in a coöperative way with other organs. Thus while the essentials are concomitant in every particle of matter they are also concomitant in every cell, in every organ, and in every body. If the expression may be permitted, nature reasons as men reason, abstractly, but is always cognizant that abstractions can be realized only in the concrete. Thus the mouth is one of the organs of the digestive system; but it also has ancillary organs of circulation, motility, reproduction and conception. The eye is an organ of the conceiving apparatus, but it has ancillary organs of digestion, circulation, motility, and probably of reproduction. The animal itself is an organ in a society of animals. Society is the culmination of a hierarchy of organs of lower grade, and every organ in every grade of the hierarchy has ancillary organs. Without entering into these subjects at length, we must give a description of these organs and functions of the animal with such elaboration only as our present purpose demands.

II

Again in this higher realm of relativity we are forced to consider the numerical relations of ultimate particles in a hierarchy of molecules which appear in kinds of substances. For present purposes we may not delay the argument for the purpose of setting forth the metabolic processes of digestion and excretion by which vegetal food is wrought into animal bodies in all the kinds of animate things; we may simply illustrate the facts necessary for this argument as they are derived from the higher animals. Digestion begins with mastication and a special substance is developed in the salivary glands to elaborate the food. Then the food is carried to the stomach, where another special substance is furnished by the liver. Finally the materials of the food are digested, excluding such indigestible substances as are taken into the stomach, and the selected and prepared food is the blood, which bears a relation to the animal analogous to that which protoplasm does to the plant. Out of the blood all of the tissues are wrought, each in its kind, and every tissue is a kind of its own, and there are kinds of kinds, so that the animal organism is a chemical laboratory engaged during the existence of the animal in building up more complex substances and tearing them down into more specialized substances, and this is metabolism, or zoöchemistry. When the animal dies decay supervenes as a chemical process. The metabolic organs, therefore, are the organs of digestion which prepare the food for the blood, the organs of secretion which furnish material to aid digestion, and the organs of excretion. The science of the chemistry of animate substances is yet in its infancy and the kinds appearing in the animate realm are at the present stage of research vicariously represented by forms. We must therefore consider them as factors of morphology. The blood is composed of serum, which is the vehicle of transportation. In this serum there float erythrocytes or red corpuscles, which are unicellular organisms into which much of the food has been converted and which is the material for reconstruction. Thus the tissues of the animal are reconstructed out of unicellular organisms. In the blood there are also leucocytes and other unicellular organisms. We cannot enter into a discussion of the functions which these additional organisms perform, but go on to remark that the red corpuscles are built into the tissues of the animal or stored temporarily in fatty structures which are subsequently used in the tissues. In so far as these red corpuscles are incorporated into the tissues by molecular rearrangement, and in so far as they are decorporated by molecular arrangement, we have metabolism; while in so far as this produces a change of form, reconstruction is involved. Here the rearrangement of molecules by number becomes structural arrangement in form, for in a body kinds and forms are concomitant.

III

The blood prepared by the organs of metabolism is delivered to the organs of reconstruction. These are the blood-vessels, consisting of the heart, veins, arteries, and capillaries, by which the material is transported and distributed to the parts where reconstruction is carried on. Thus there is a system of organs for reconstruction.

That which we found in the geonomic realm as spheres and in the phytomic realm as blasts, we here find in the zoönomic realm as derms, and we have the ectoderm, esoderm, and endoderm as encapsulating bodies, with a concentric nucleus. These cells are modified as they are combined into larger cells, but the cellular structure is still preserved in organ and individual. The metabolic organs or those of digestion, secretion and excretion are compound nuclei inclosed in cellular sacs; sometimes these sacs are greatly elongated so as to be tubular, but in general the organs of digestion and excretion have a cellular form with permanent compound nuclei or with passing nuclei when they are conduits to contents.

In the circulatory system of organs the same dermal structure is observed with its triune elements. In the heart there is a compound nucleus, but in the artery or vein the nucleus is passing content, and in the higher animals there is a vast system of ramifying tubes, which are duplicated as arteries and veins directly connected in the heart, and functionally connected with the capillaries.

In the activital or muscular system every organ is a fascicle of muscles, and each member of the fascicle has a dermal structure. The nucleus of the heart is a compound muscular organ of this character, whose function is to impel the blood; muscular tissue undergoes important metamorphoses, becoming tendonous and osseous for a variety of mechanical purposes. Tendons are dermal in structure, and bones are sacs enclosing nuclei of osseous tissue.

It was in the bony structure that homologies of form were first discovered, and the homologies of the vertebrate skeleton was at one time the sole theme of morphology. Of especial interest were the transformations that were discovered in the vertebræ in the development of limbs and cranium; but the subject of morphology has passed out of this stage into a wider field embracing all realms of nature. Only of late has it appeared in the morphology of formations and land features.

The reproductive cells are compounded into organs still preserving the typical structure.

It is in the organs of sense that the most marvelous changes of form are discovered. The metabolic sense organs are thrown into two not thoroughly differentiated groups known as the sense of taste and smell; but these groups seem to be continuous, that is, without a well-marked plane of separation; the one group, that of taste, taking cognizance of liquids, the other, that of smell, taking cognizance of vapors. The organs of touch are distributed throughout the skin; these are primarily the sense organs of form. The sense of stress or pressure seems to be in or immediately under the skin; the sense of duration or time is the sense of hearing, and the sense of ideation is the sense of seeing. The homologies of mouth and nose, skin, muscle, ear and eye, are yet imperfectly known; though much research has been bestowed upon them they are difficult to understand. Thus there are homologies of form in all the hierarchy of organs, for they all have the dermal structure.

IV

There are five modes of motility called functions; these are the functions of the metabolic, circulatory, muscular, reproductive and reasoning organs, as heretofore set forth. Metabolism continues as long as animate life continues, but is increased when the special function of the organ is stimulated; that is, both anabolism and catabolism increase in the organ by increase of its special function, but metabolism wanes as special function wanes.

The reasoning function may increase or retard the

other functions, though it cannot wholly inhibit their action nor can it increase their action beyond certain limits. This fact is well known to psychologists and physiologists. It seems to be accomplished by the promotion of metabolism.

Here we are confronted with a problem met before concerning the nature of affinity which we have not been able to solve. If it were permitted to hold the doctrine which has been entertained by some great minds that every particle of matter has judgment, the question would be solved and affinity would be conscious choice. Affinity is often expressed as choice and many chemists have held this doctrine.

Next we have to consider how molar motion in the individual is self-directed. We have seen that molar motion is accomplished by compound organs. These organs are found in pairs, so that one acts against the other. We have seen, too, that the mind can accelerate metabolism and the mind can direct the motion of the animal. Now let us suppose that the mind can accelerate anabolism in one muscle and catabolism in its opposing muscle, and we have a very simple explanation of the nature of the self-direction of muscular energy—the nature of the mechanism by which the animal may walk to the east or west at will. That muscles are in pairs is an anatomical fact, and that the one contracts while the other relaxes is a physiological fact, and that the mind somehow controls this muscular activity at will is a psychologic fact, and the whole thing is rendered simple and clear by the doctrine that anabolism in one muscle and catabolism in its opponent are each under the control of mind. But the mind of the cortex does not consciously choose the association of the several particles involved in metabolism. The affinity which is involved in metabolism must be the choice of the particles themselves, in obedience to commands issued by the organism of unicellular particles of which the body is composed, these ultimately acting in obedience to the command of the cortical consciousness. Metabolism is controlled by the central mind in some manner or other. Believing this we must infer that the particles of the muscles are conscious as units in a hierarchy of organs which at the other pole is the cortical consciousness. Here we first reach the facts the explanation of which seems to require the hypothesis that consciousness primarily inheres in the ultimate particle. If this hypothesis is accepted, we have the fundamental doctrine of psychology.

Science has demonstrated that motion cannot be created or destroyed. Mind, therefore, cannot create motion but only direct it. Mind directs the motion of the body by directing the motion of the organs of locomotion, and these are directed by the device of opposing muscles—the one being contracted and the other relaxed. So the choice of the animal is delegated to the choice of the organ, and the choice of the organ is delegated to the choice of the muscles. The muscles, therefore, must have the power of choice, which it also delegates to molecules. Therefore the molecules must have choice. We know that every unicellular organism of the blood is an independent animate being, with consciousness and choice. These independent animate beings are incorporated in the tissues of the animal having self-activity. We must therefore suppose that they retain their choice and consciousness, and the same choice seems to be exercised by every particle of the molecule; if so, animate existence as consciousness and choice is universal in every particle of matter.

The human body is a hierarchy of conscious bodies. In this hierarchy the lower members are controlled by the higher members. The lowest members are ultimate particles and the highest member is the cortical body. Now the cortical body controls all the others in the hierarchy and it ought to receive intelligence from all the others, for the consciousness of the particle is transmitted to the cortex, and the will of the cortex is transmitted to the cortical body, but only those which require regulation by it. Not all of the judgments of the cortical body, but only those of the particles which need regulation in a particular part, are transmitted to special particles. The government of the human body in all its hierarchy of bodies is strictly analogous to the government of a nation where the governing body of the nation is not cognizant of all which the individuals do, but it receives intelligence about the way they do in respect to those things which it attempts to control and it controls the individual only in those actions which are necessary to the welfare of the body politic. Thus the cognition and volition of the controlling body is but partial. There is local consciousness and local self-government. We will find some confirmation of this doctrine as we proceed, but its final elaboration will be more fully made in a subsequent work. Stated in our own terms, this is the doctrine of modern scientific physiology and psychology,

V

As in the geonomic realm so here in the animate realm there are processes. As in the phytonic realm so in this there are generations; now causation appears under a new aspect as development. The animal is composed of organs and these organs develop as they are exercised under the stimulus of mind, for while they are coöperative one part of a system may be developed at the expense of another, so that one organ in a congeries of organs may have great development while another organ in the same congeries may be neglected and ultimately in a series of generations may become atrophied. There is a law which finds its chief expression in this realm where one organ of the same system may be developed, while another may be atrophied. This may be stated as follows: progress in unification in organs of the same function is progress in rank. There is another law, the correlative of this; it is that the differentiation of functions with distinct organs is progress in rank.

The mechanical causes of force, form, and kind are conditions that are genetic, while the conditions of conception are teleologic. The teleologic conditions are concomitant with the genetic conditions.

VI

It seems probable that every particle of matter has consciousness and choice; certain it is that every particle of animate matter has these properties. In the animal body all of the particles coöperate and for this purpose a special nervous system is provided. In this system there is a congeries of cells, whose function is conception, connected by another congeries whose function is association. The conceiving cells are ganglia, the associating cells are medullary or fibrous. A group of such gray cells is connected with other groups by white fibers, and finally all of the ganglia are connected with all other animate cells of the individual by fibers. Thus the nervous system is a congeries of ganglionic organs, connected with and presiding over the other systems of organs. The fibers are connecting lines between the outer systems of organs and the special ganglia of the organs. These ganglia are grouped in the hierarchy of nervous organs by intervening fibrous nerves until they reach the master ganglion of the brain, which is the cortex. There is a peculiarity of the nervous system in the relation between the cells of the ganglia and the fibers of the connecting nerves, in that the fascicles of fibers are not structurally continuous with the ganglionic cells. Thus when a feeling starts in the end organ and is produced by its activity, it is carried along the fibers through the hierarchy of ganglia to the central cortex; the intervening ganglia may continue its transmission to the cortex or, as it seems, may inhibit it; or when, as in a dream, the system is relaxed, the impulse may go astray among the cells of a ganglion, and may be transmitted by unwonted fibers to the cortex at some incongruous point, for the cells of the ganglion constitute a shunting or directive apparatus by which impulses from one region are directed to others throughout the system. Now all of the metabolic, circulatory, motor, and reproductive organs are themselves organs for the initiation of impulses to the nervous system, and the ganglia of this nervous system, especially the cortex, are organs for the initiation of impulses that are conducted by the fibrous nerves to the metabolic, circulatory, motor, and reproductive organs. A ganglion seems to have the power to distribute these impulses to such point in the peripheral organs as they may select, but the central ganglion or cortex cannot directly reach the peripheral organ, but only through the intermediate ganglia in the hierarchy. An impulse emanating in the cortex is delivered to its nearest ganglion in the line in which it should go; this ganglion in turn directs it to another or to any group of end organs. Thus all of the systems of congeries of organs of which the body is composed are put in relation to the cortex. An impulse which originates in any organ of the complex system when transmitted to a ganglion I call a feeling impression.

Having seen the nature of the apparatus by which the other organs are put into communication with the ganglionic organs, and finally with the cortex through feeling impressions, it becomes necessary to exhibit the apparatus of the nervous organism which exists to connect the cortex and subordinate ganglia with the world external to the periphery of the body.

This apparatus consists in the sense organs and the fibrous nerves by which they are connected with the cortex. For the sense of taste and smell, which are metabolic, we have two organs that are not very well differentiated in structure, nor are they well differentiated in function, although they seem to be more thoroughly differentiated by the nature of the stimuli; for taste the object must be reduced to the fluid state, and for smell it must be reduced to the vapor state. Both of these organs have their nervous bodies connected by fibers with the central ganglia. The mouth and the nose are simple organs for the accumulation of sense stimuli, single in the one case and partially double in the other, but the nervous organs to which they lead and which they unify are many.

In the skin-covering of the body there are many tactual organs, which are unified through the continuity of the skin itself, yet they seem to be disparate not only in organ but in function. They are also connected with the cortex, but through ancillary ganglia, which are themselves ancillary brains. Touch is the primary organ of form.

There also seem to be organs of pressure either in the skin or immediately beneath it, though they have not been clearly made out. The fibers of the muscles themselves may be the end organs of the motor system, and it may be that nerve fibers everywhere accompany muscular fibers. Thus we know that the motor system is connected usually through ancillary ganglia with the cortex. The end organs of this system, be they the muscles themselves or specialized parts of them, are the organs for conveying to the cortex impressions of muscular force.

For the sense of hearing there are two organs for gathering the impulses which are propagated through the atmosphere, but in each there are many nerve organs. They are also connected with the cortex by their fibers. The semicircular canals seem in man to convey only feelings, but in aquatic animals it is probable that they are true sense organs, and convey sense impressions brought to them through the medium of the water. The ear is the primordial or fundamental sense by which time is conveyed to the cortex.

The eye is the organ for conveying sense impressions that are received from objects at a distance through the medium of the ether. Primarily or fundamentally it is the organ by which the conscious movements of other bodies are conveyed to the cortex.

In man and probably in many of the lower animals all of these senses are highly vicarious. This is pre-eminently the case with the eye. This organ, by reason of its self-activity, is peculiarly adapted to a great variety of vicarious functions, for it can adjust itself to direction through its muscles or by accommodation to distances and degrees of light. The faculty by which the eye moves and accommodates itself, together with the rapid vibration of ether particles, renders it possible to receive many sense impressions which come to it with a speed which is for all practical purposes instantaneous. For these reasons and for others that hereafter will be set forth, the eye is a universal organ of sense impression.

The ear also is highly adapted to vicarious functions, the air being the medium whose vibrations are rapid, though to a less degree than those of the ether. In the early history of mankind, when language was chiefly oral speech, the ear was rapidly developed in vicarious functions, especially in the function of conveying the properties of mind observed in other human beings, for by this organ men learn that other human beings have ideas and emotions like their own.

The motor sense also seems capable of becoming highly vicarious, for those persons who are deprived of sight and hearing can yet through the aid of this sense obtain a knowledge of the world which they can neither see nor hear; and what is more wonderful still, they can yet gain a knowledge of the ideas and emotions of their fellow men. The other senses in a still lower degree are vicarious.

It will be seen that I do not consider the temperature feeling to be a sense or to have sense organs. The temperature feeling seems to be the feeling of the functions of the circulatory system in degrees when it partially congeals the blood, or increases its fluidity, and is a feeling like that of a burn when it injures the skin. The distinction which is made between a feeling impression and sense impression is fundamental, and must be considered when hereafter the nature of cognition is discussed.

VII

Essentials are comprehended in the same particle, and are thus concomitant, and related in different particles, and are thus correlative. As particle is related to particle, so unit is related to unit, extension to extension, speed to speed, and persistence to persistence. Now we have discovered another property in bodies, which we have found in inanimate bodies as affinity or choice, and in animate bodies as consciousness and choice. There can be no choice without consciousness. Consciousness is to choice what unity is to plurality, what extension is to position, what speed is to path, and what persistence is to change; that is, consciousness is the absolute, choice is the relative.

Thus for every absolute we find a relative; for every constant a variable. Unity as an absolute has plurality for its relative; extension as an absolute has position for its relative; speed as an absolute has path as its relative; persistence as an absolute has change for its relative, and consciousness as an absolute has choice for its relative.

Unity and plurality constitute number, the unity being absolute and constant, while plurality is related and variable; this is the fundamental definition of number.

Space is composed of extension and position, the extension being absolute and constant, the position relative and variable. This is the fundamental definition of space.

Motion is speed and path, the speed being absolute and constant, the change relative and variable; this is the fundamental definition of motion.

Time is persistence and change, the persistence being absolute and constant, the change relative and variable; this is the fundamental definition of time.

Judgment is consciousness and choice, the consciousness being absolute and the choice relative; this is the fundamental definition of judgment.

Let us further consider these properties to bring out another phase of the subject. Unity is the substrate, foundation, ground or condition of plurality, for without units there can be no pluralities. Unity, therefore, is independent of plurality, but plurality is dependent on unity. There are many particles that have extension or space occupancy; thus there are many positions. Extension is the substrate, foundation or ground of position, for the several positions depend on the several units having extensions that exclude one another in the occupancy of space. Speed is the substrate, foundation, or ground of path, for every speed produces a path, or in other terms, every path is dependent on a speed. Every unit having extension and speed has persistent duration; but as these units change in position and also change in trajectory, they could not change if there were not something that persisted through change. Persistence, therefore, is the substrate, foundation or ground of change. Consciousness is the substrate or ground of choice, for if there is no consciousness there can be no choice. Thus it is that in every one of the properties there is a substrate or a support and that which is supported, or, in other terms, a ground and that which is grounded, or in still other terms, a foundation and that which is founded, and finally an independent and a dependent. This is but another way of saying that in every one of the properties there is a substrate and a dependent. The substrates are unity, dimension, speed, persistence and consciousness; the dependents are plurality, position, path, change, and choice.

It will be seen that we can call a particle a unit or we may call it an extension, or a speed, or a persistence, or a consciousness, and these several names refer to the same particle because it has the five concomitant essentials.

In the foregoing presentation the nature of the properties has been deduced from knowledge, with which every intelligent person is possessed, and which rests upon the experience of the race. No recondite induction or deduction has been necessary, but only the statement of known facts in proper sequence has been required to understand the nature of the five properties, except in the case of judgment, which is made analogous by hypothesis.

This is the result at which we have arrived in the foregoing discussion.

One particle by itself has unity, extension, motion, persistence, and if animate, judgment; but by reason of others it has plurality, position, path, change and choice. What it has by itself we call its essential concomitants; what it has by reason of others we call its relations. Concomitants with relations we call properties, and as the essentials are concomitant the properties are concomitant; hence the number cannot be absorbed by one, the space by a second, the motion by a third, the time by a fourth, and the judgment by a fifth. Properties, then, are concomitant and relational.

The theory of hylozoism, which I have presented in this chapter, is very old, and has had many illustrious champions. When alchemy was developed into chemistry a great impetus was given to it. The discovery by Darwin and the masterly advocacy of evolution by Spencer, through which the doctrine of the survival of the fittest was established, for a time gave a decided check to the theory. The blow struck by Spencer was especially efficient, for Spencer resolved all of the properties into force with a clearness which left no room to doubt his meaning.

A host of scientific men following Darwin and accepting the doctrine of the survival of the fittest have found it to be inadequate as a single theory of evolution. There are other laws, especially one expounded by Lamarck. I myself have set forth a new doctrine of evolution as that of culture, and in a subsequent chapter of this work I shall set forth the doctrine of evolution in which I shall attempt to prove that the fundamental law of evolution is the law of affinity by which bodies are incorporated, and hence that evolution is primarily telic.

In the five fundamental realms of nature, ethereal particles are numerically related and numbers are organized. Stellar particles are related in numbers and forms, and forms are organized. In geonomic bodies forces as well as forms and kinds are organized. In plants causations are organized as generations as well as forces and forms and kinds. In animals concepts are organized as well as causations, forces, forms and kinds. In every one of these systems there is a special differentiation and integration of organs; so the entire body is organized in a hierarchy of organs. This may be stated in another way. In ethereal bodies, which are probably ultimate particles themselves, numbers are organized. In the stars numbers and spaces are organized. In the geonomic bodies numbers, spaces, and motions are organized. In plants numbers, spaces, motions and times are organized. In animals numbers, spaces, motions, times, and judgments are organized. Or again, it may be stated in another way. In ethereal bodies units are organized. In stellar bodies units and extensions are organized. In geonomic bodies units, extensions and speeds are organized. In plants units, extensions, speeds and persistences are organized. In animals units, extensions, speeds, persistences and the consciousness of many particles are organized. While every particle in the universe has consciousness and choice and hence judgment, it is only in animals that we find judgments organized as concepts. Only animals have reason.

The various doctrines of hylozoism heretofore presented in the history of philosophy, consciousness and reason have been confounded. The terms mind and reason are nearly synonymous. Reasoning is a process, as we shall hereafter show, and mind is that which reasons. Thus these two terms refer to the same thing, the one when it is considered as a process of an organism, the other considering it as an organism. Reason is a function of animal organism. Every particle has consciousness, only animals have reason.

CHAPTER VIII

QUALITIES

There is another class of relations which here require careful consideration. They will be called qualities. Sometimes the words property and quality have been considered synonymous, while the words quality and class or category are often used as synonyms. Perhaps the distinction now made between properties and qualities has never been set forth. I think that the foregoing chapters will have made clear to the reader the sense in which the term property has been used. These properties in the five realms of nature, namely, the ether, the stars, the rocks, the plants, and the animals, all subserve human ends or purposes, which may be considered as good or evil. In this manner qualities arise, while terms denoting these qualities are found in all languages. These quality terms have the characteristic of being more or less vague, in that they may instantly change with the point of view. Some illustrations will be given to make this distinction plain. Number is a property. Here are five apples and the number cannot be changed without adding or substracting therefrom, but the five apples may be few or many by a change in the point of view. Five apples in a tray at a dinner board where twelve persons are sitting are few, but upon the plate of one of the guests are many. Thus it is that a number may become few or many by some circumstance or purpose in view. Few are thus qualities, while five is a property. A barrel of apples on a table would be many or very many; in the cellar plenty; in the warehouse when the steamer is seeking a cargo it would be few, and the merchant would not be considered untruthful if by a figure of speech he affirmed that he had none.

Again, extension and form are properties, but they may easily become qualities where there is some purpose in view. A pin may be large or small in relation to the hole which it is to fill in the timbers of a house; the same pin may be too large for one purpose and too small for another. The watchmaker uses a pin so small that it can be seen only with care, and yet it may be large or too large for the purpose intended. A hill in the Park Mountains would be called a mountain in the Catskills, and a mountain in the Park would be called a hill in the Himalayas. Thus properties are transformed into qualities by ideal circumstances.

The railway train is fast to the man who is driving an ox-team, but the train is slow to the mother who is on her way to the death-bed of her child. An old man may say at one moment that the day is long, and in the next that life is short. To the laborer who is bent on his task the hum of the machinery is scarcely heard, but on his couch at night the tick of his clock is loud. The razor is beautiful and good in the hand of the skilful barber, but it is ugly and dangerous in the hands of an assassin; thus properties are transmuted into qualities by human ideas. Red is beautiful in the rose, ugly in the spot of blood on the floor. The sheen of sable in the ousel is beautiful, but the sheen of sable on the carrion-loving buzzard is ugly. If all serpents were harmless, gentle and intelligent, their lithe forms and gliding motions would be beautiful. If robins were poisonous their red breasts would be symbols of horror. If the red lightning and the crimson cloud could change relations to men's ideas of good and evil, the one as the harbinger of summer rain and the other as a visit of death, the lightning would be a thing of beauty and the cloud a terror.

The coming of the rain may be welcomed by the husbandman who has planted his field of corn; it may be unwelcome to the belated traveler. Time is long and weary to the invalid on the couch of pain; time is short and joyous to the child in the park.

It is thus that properties become qualities through our ideals, through the purposes which we have in view. There is no difficulty in distinguishing between qualities and properties as they have here been defined. Properties are not qualities and qualities are not properties, but qualities are founded upon properties. Properties are qualities when they are considered teleologically. It is right, therefore, to say that properties are real in the sense that they are grounded on matter and that qualities are ideal in the sense that they are dependent for their existence upon the mind. When we reflect upon these facts nothing can be more simple. The distinction can be discovered without difficulty and it would seem that there need be no confusion between properties and qualities as here defined. To affirm properties is to affirm inseparable concomitants of matter, but to affirm qualities is to affirm things that change with the point of view. I see a man suddenly push another upon the street, and think it rude, and am indignant. The next moment I see that he saved him from falling into a pit, and in an instant the quality of the act is changed, and I call it wise and kind, while the activity as property remains the same.

From the days of Aristotle to the last book of philosophy, substance and the properties of which it is composed, bodies as compounded substance and hence compounded properties, relations and compounded relations, qualities, and compounded qualities all have been under discussion, and attempts have been made to define them.

These distinctions, which seem simple and are simple when understood, and may be understood by every intelligent man, have led to tomes and libraries of discussion and disputation not always friendly and charitable. There are those who affirm that qualities and properties are all one as ideal; there are those who affirm that qualities and properties are all one as real or material. And thus we have an idealistic philosophy and a materialistic philosophy. A few idealists have gone so far as to affirm that not only qualities but properties, bodies and relations are ideal; that there is no material or real world which exists except as it is created by the mind and that all these things exist only in mind.

The difference between qualities and properties was vaguely seen by Aristotle, but seems to have been unrecognized by Plato. In modern times we find Locke, with a clearness never before exhibited, giving the distinction between properties and qualities, though he called them all qualities, but the names used are of little moment. He divided qualities into primary and secondary; what are here called properties he called secondary qualities. But at his time the nature of force was unknown and the laws of evolution or time were undiscovered and many of the properties of force and change were relegated to his second class and confounded with what are here called qualities. Then he added a third class which he called powers; so the properties of force were divided between secondary qualities and powers. Dropping his term as primary and secondary qualities, and using the terms properties and qualities in their stead, it is proposed briefly to explain the errors into which Locke fell. In his time his errors were excusable; at the present time they are inexcusable. All of this can now be set forth and the truth demonstrated as simply and clearly as a proposition in Euclid, and it must be understood if modern science is to be understood, for upon these simple, self-evident propositions all modern science is founded. Since Locke all later writers, so far as my reading extends, instead of clearing away Locke's errors have piled up a mountain of new fallacies. To reduce these questions to their simple elements it becomes necessary to go back to Locke.

The correlation of forces which has its ground in the persistence of motion was unknown in Locke's time, though Locke himself affirmed it. In his discussion he clearly set forth that numbers are primary qualities—*i.e.*, properties; but he does not see that kinds are derived from number and also are properties. He clearly explains that extension and all the properties of form derived therefrom are properties. He clearly sees that motions are properties, but he does not see the relation between motions and forces, so he places some of the forces in the second class of qualities and thus includes them in what we call qualities, while others he includes among powers. Thus classes, forces and durations were practically left in the second class and among powers. The nature of the first class he clearly understood and explained, and finally he refers the second class of qualities and powers also to a foundation or substrate in qualities of the first class, or properties. His second class of qualities he included with pains and pleasures, which are true qualities. He clearly saw that good and evil, however expressed as pleasures, satisfactions, joys and delights, or as pains, discomforts, dangers and horrors, formed another class of attributes. But with them he grouped classes, though he does not make this plain; but he does make it plain that he grouped many forces and many changes in his second class of qualities.

Since Locke's time this classification has been modified mainly in the direction of his errors. More and more have properties been considered as qualities, and a school of idealists has sprung up who hold that all properties are qualities in the sense in which these terms are here used. At the same time a school of realists has sprung up who hold that there are no qualities, but only properties, as these terms are here used. By what course of reasoning did Locke lapse into error? On carefully examining this matter it will be seen that while he did not discuss the whole question fully and left much unsaid that should have been said, he clearly understood his position; yet it will be seen that he stumbles over those properties of force that are revealed to us through the senses of seeing, hearing, and smelling. He clearly saw that the bodies revealed to us through these senses do not act directly as bodies upon the self, but in the case of seeing and hearing through media and in the case of smelling through the action of minute particles dissevered from the bodies. At least all this may be justly gathered from his statement, though he is not always clear upon these points. It is fair to Locke to credit him with this degree of insight into the truth. He believed that in seeing there must be a medium between the body perceived and the perceiving mind, but he did not clearly understand it as the universal ether. In his time the existence of the universal ether was a doubtful doctrine in the history of science. Locke denied the validity of the *actio in distans* in his first publications, and he never retracted, but under the influence of the supposed opinions of Newton in regard to the attraction of gravity, Locke affirmed that he was not prepared to assert that God could not do things in any way he pleased. Had he known what we now know, that Newton used the term attraction in a metaphoric sense, and no more believed in *actio in distans* than did Locke himself, he would not have made this apparent concession to the opinions of Newton.

It still remains, however, that Locke believed and taught that certain properties of force (especially those manifesting themselves to the senses above mentioned) and many properties of change are qualities and do not exist as properties or primary qualities. Fallacies of force and change were still current in his time, for the correlation of forces through the persistence of motion was unknown and untaught, and the fallacies of evolution were yet to be dispelled. This state of things has passed away, and no man who now understands light or heat will call it a quality in the sense in which the term is here used, but a property inherent in matter itself. At first view it seems strange that Locke fell into this error in the case of sound, but it must be remembered that in his time the kinetics of gas was unknown, and although Locke and his predecessors for two thousand years had understood that sound was a mode of motion, yet it was very vaguely or inadequately explained.

Locke's contemporaries and successors have but added to the confusion in which the subject was left by himself. Spencer takes up this subject for discussion in three chapters of his *Psychology* under the subject of static, dynamic, and statico-dynamic attributes. We first note that he replaced Locke's term of qualities by another, namely, attributes. He did not discuss Locke's classification, but that of Hamilton, which is much more vague than that of Locke, but Hamilton, like others, had introduced a third class between the primary and the secondary, which was called secundo-primary. Spencer adopted this threefold classification, but used the terms static, dynamic, and statico-dynamic. It will be remembered that Spencer was a Monist, and believed that the primordial unity is based on dynamics or reified force. With him all the properties, and in them he included qualities, manifest only the primordial force. This was his first error. His second error was to neglect number and to consider class as classification, or a process of the mind, and not a property of bodies discovered by the mind. Then he presented his two classes, one based on dynamics and the other on statics, but statics is not the other to dynamics, but the other to change; state and change are the reciprocals of time. The reciprocals of force are action and passion or action and reaction. You may read Spencer on this subject with great care many times, as I have, and you will see that he himself is vaguely conscious of this illogical proceeding and affirms that he uses the term statics with an especial meaning devised for his own purpose; but under dynamics he appears to include change, although he purports to be the philosopher of evolution, and under statics he includes a part of the properties of duration and change and a part of the properties of number and class and of extension and form. It is thus that the confusion introduced by Locke in his discussion, due to the ignorance of his time, was still further increased by Spencer, and his three chapters on the attributes of matter constitute a monument of errors. An erroneous classification is the bane of science, for it throws phenomena into false relations and makes that which is simple appear to be complex, difficult, profound and even unknowable, as Spencer believed.

Locke's "Essay" introduced a new theme into philosophy, which at last comes down to us in the form of epistomology. It seeks to discuss the activities of mind and the certitudes of its conclusions. Berkeley seized upon Locke's explanation of vision and amplified it. Neither Locke nor Berkeley clearly saw that the properties of bodies discovered by the several senses are integrated by conception in such a manner that one sense impression becomes a symbol or mark of all the properties belonging to the body which are known to the mind; that a light impression, a sound impression, a taste impression or a smelling impression are by conception transformed into symbols of the body perceived with all its properties. Failing to understand this in its full significance, and science not having explained the nature of light, heat and other forces, all forces were by Berkeley considered to be qualities as the term is here used, and then he made a further step, that all properties are but qualities, and have their existence only in the mind. Thus it was that Berkeley robbed us of the beautiful world, but with a literary skill that is alluring; he was not a vulgar highwayman crying, "Stand and deliver!" but a knight of the green wood who courteously invoked our assistance in yielding to him our treasures.

Hume took up the same problem and with sturdy blows destroyed the world, and reason was crushed in its fall. Then in Germany Kant, Schelling, Fichte and Hegel essayed to solve these problems; Kant leaving behind a monument of criticism erected into antinomies where truth and certitude are lost. Fichte carried the whole subject to its logical conclusion by reducing it to an absurdity. It was a simple demonstration the meaning of which he never knew, dying in a mist of reification. Hegel, seeing the contradictions of Kant and Fichte and accepting their conclusions, developed the most elaborate and artificial philosophy ever presented in the history of human thought—a philosophy of contradiction, a scheme of the negative by which it was attempted to show that words are divine, but the world is finite and contradictory, and that every proposition affirmed of the world contains within itself its own contradiction, and that words must be believed and that sensation, perception, understanding, and reflection create phantasms.

So these problems have come down to us. In the meantime an army of scientific men have been at work clearing away the fallacies of imperfect reason by designed and skilful investigation. Mysterious forces have been resolved into their simple elements as the motion of matter in collision, and the metageneses of the world have been resolved, and the laws of evolution formulated, and the subject is once more taken up by Spencer with a literary skill equal to that of Berkeley or Plato, and with the powers of an advocate never excelled. The attributes or things which may be attributed to an object are properties and qualities. It was the distinction between properties and qualities that the Greeks sought to characterize as noumena and phenomena. Noumena are the properties of bodies as they are in themselves, while phenomena are the qualities of bodies and the fallacies which we entertain concerning them. But when in later times noumena were held to be occult or mysterious substrates, then science adopted the term phenomena as synonymous with properties.

Qualities give rise to emotions, for qualities are good and evil. All properties may be considered as good or evil in relation to man's wants. The emotions are founded upon the cognition of good and evil. We are not in this volume to set forth the good and evil of environment, nor their cognition as emotions. All of this subject must be treated in a subsequent volume. In this volume we are endeavoring to explain, first, what are properties and bodies, and how they are cognized. This brief reference to the cognition of qualities must here suffice.

CHAPTER IX

CLASSIFICATION

The science of number is natural, for units and pluralities are found in nature, but measure is conventional, for conventional units of measure are used in order that undiscovered numbers may be represented by their equivalents in computation, for while we may not be able to discover the number of natural units in a body we may be able to measure its form in conventional units of extension, and for some purposes of computation these units serve the desired purpose.

There are other computations which are not properly subserved by the measurement of form. Here we measure the force which the body exerts through the action of gravity and determine its mass in units of weight, and these mass units serve the same purpose in our computations that higher units of number would serve if we were able to count the particles. Thus the science of number is natural, but the device of measure is conventional. It serves a useful purpose in that it enables us to represent by numbers certain facts about bodies which we are not able to discover as natural numbers by reason of their multiplicity and minuteness; so we assume that one concomitant property represents the others. This we measure. We do not search with the microscope for atoms and count them, but we consider their forms as extensions or their forces as masses and reason about the artificial numbers derived therefrom by measurement with the same degree of certainty that we would have if we should actually count the particles. Thus measure is devised in order that we may consider numbers when the actual numbers are concealed from observation. That every property is concomitant with all others is thus assumed as the fundamental doctrine of mathematics where quantitative reasoning is held to be exact and irrefragable. All this depends upon the law that the essentials are persistent in the particle.

While measure is thus conventional there is still another conventional usage in the science of mathematics. In natural units bodies are the higher units of particles, the particle and the body are units of different orders, and the different orders of units in nature are thus coextensive with all the bodies of the universe. Thus there is an infinite system of orders of numbers; but man devises a numerical system where a definite plurality is considered as a higher unity, and such a system serves him a valuable purpose as a labor-saving device for the mental faculties. He cannot stretch his mind to the concepts of natural units of particles in natural higher units of bodies, but he creates a representative system, so that the multiplicities of nature, which are infinite, may be representatively considered by the finite mind.

In conventional number the units of different orders are compounded symmetrically in constant ratios. Early in the history of language, while it was largely gesture speech, the fingers of one or both hands or the fingers and toes were used as an abacus by which numbers were told off; and this led to a habit which has continued and developed so that in the various languages of the world it is found that the number five, the number ten or the number twenty has been used as the normal ratio between conventional orders. Of the three methods the decimal has been retained in civilization as the one used in enumeration, computation and notation. By this device a plurality of units are arranged in a system of orders, ten units constituting the first order, ten of these the second, etc. In this manner numbers are classified as kinds in series for the purpose of convenient counting. Counting is a compound process of two coördinate elements; one determines the kind, the other the series, and determination of kind logically precedes enumeration. The kind must first be determined and then seriated. The kinds may be natural or conventional, one or both, and the series may be natural or conventional, one or both. When we count horses in the field we count a natural kind, but we seriate only those in the field as a conventional series. We must not confound horses with stumps if we are to get a valid sum. We may place stones, blocks of wood and fragments of paper as marks of sites where trees are to be planted, but we classify them not as stones, blocks of wood, and fragments of paper, but as marks. In this case the kinds are conventional. Conventional counting and classification differ in this respect only that in counting the series is conventional, while in classification the series is natural. In counting the all of the kind is the all of our purpose; in classification the all is the all of nature. Then we must remember that in mathematics, number is taken as the representative of the other concomitant properties of quantity and that they are reduced to number by measurement, while in classification kinds are used to represent the other properties and they are reduced to kinds by logical convention. While in conventional counting we consider kinds in series, so in classifying the bodies and properties of nature we are compelled to consider kinds in series.

It was more than a chance that produced the decimal system, for the universe is pentalogic, as all of the fundamental series discovered in nature are pentalogic by reason of the five concomitant properties. The origin of the decimal system was the recognition by primitive man of the reciprocal pentalogic systems involved in the two hands of the human body, and the pentalogic properties are always in pairs. While the properties are five, they are manifested in reciprocal pairs.

The universe is not an endless series of infinitesimal variables, but it is a universe of divergent series which spring from an ascending series as branches spring from a trunk. In the branches the extreme variation appears in the extremities of the divergent branches, but the branches are not linked to one another by these peripheral extremities but by their trunk connections, and the grand advance in nature is made as an ascending series as by a trunk.

When we study a group of plants or animals that are intimately related, as, for example, the members of an order, and compare them with the members of another order, the two orders are found related not by their highest members but by their lowest. It is thus that two branches of phytonomic or zoönomic species are found related to each other by discovering the synthetic form which belonged to the ascending or trunk series.

Synthetic forms are often extirpated by time, and to a large extent living species are found in well-demarcated groups, this demarcation being the clearer by reason of the extirpation of the synthetic types of the trunk, while the branch groups divergently elongate until an extreme differentiation is found. Sometimes whole branches are extirpated and thus are found as fossils. Species multiply by the splitting of branches and each new branch consitutes a lineal series of individuals which are separated by the extirpation of the main branch; while the main branch remains the new branches are held as varieties.

The true method of classification, therefore, is not by invention but by discovery.

The growth of a mineral is a progressive change by internal metamorphosis of the molecules. The growth of the individual plant is accomplished by successive additions of particles, and is thus a serial kind, while the growth of the individual in the animal is accomplished not only by a constant addition of particles, but also by a concomitant subtraction of particles; the individual is doubly a serial kind.

A species is a series of connected individuals differing from one another by minute distinctions but differing from other species by gaps; such a group is the lowest demarcated class. A variety is an inchoate species not marked by gaps or discrete degrees. Species are further classified in hierarchies, when the species becomes one of a series of species. The production of a species is nature's method of summating a series, and a production of any higher class is still another method of more distinctly summating a series of series. Series spring from the division of trunks, and may be traced back to their origin; classification, then, becomes seriation of species in such a manner as to exhibit their origin in less differentiated species.

The kinds of nature considered in the series of nature are classes, and these are regrouped in hierarchies which are systems of classes. Every science of such a grand group of bodies gives rise to a special science and thus we have systematic mineralogy, systematic botany and systematic zoölogy.

We have seen that the other properties of a particle when treated in the science of mathematics require conversion into terms of number. Space properties are measured by conventional units, and are thus reduced to number. Motion or force properties are measured in terms of space and these again are also expressed in number. Times are measured in terms of motion, the motion in terms of space and the space reduced to terms of number. It is thus by the device of measure that all the other properties of matter are reduced to number for the purpose of verification. Abstract mathematics is therefore the science of number, but applied mathematics is the utilization of the laws of mathematics in concrete investigation by the device of measure, while chemistry is the science of natural orders of number.

Now, that which is true in the conventional science of mathematics finds its analogue in the natural sciences, for all the other properties of bodies are reduced to kinds for the purpose of logic. Forms are explained as kinds, forces as forms and then as kinds, and finally causations are reduced to forces, the forces to forms and these forms to kinds. Thus all the natural categories are reduced to kinds, as quantitative properties in mathematics are reduced to conventional numbers.

It is for practical reasons that man has reduced all other properties to numbers, for as counting can be accomplished only by classification, so properties can only be treated in mathematics when they are reduced to number by measure. Counting serves to determine the extent of a conventional group, while classification serves to determine the extent of a natural group.

Language is impossible without classification, for most words are class words. It therefore becomes necessary in the arts, both industrial and linguistic, to classify, and mankind through all the history of culture has been engaged in classification. But the reduction of the other properties to kinds does not reduce the whole of science to classification any more than the reduction of quantities to number reduces all verification to mathematics. There is still a logical verification independent of mathematical verification, and there are still forms, forces and causations to be considered, although for deductive logic it is necessary to reduce them to kinds.

Kinds as species become orders of kinds or classes, and are thus multiplied. When kinds are considered two correlates are found which cannot be expunged; likeness and unlikeness; and when considered in this manner they are classes. A fundamental likeness is discovered in all bodies, for all bodies are composed of matter.

In mathematics bodies are considered in their

quantitative properties, which are number, space, motion, time, and, in animate bodies, judgment. But in systematic science bodies are treated as categories, which are kinds, forms, forces, causations, and, in animate bodies, concepts. So, in mathematics, while quantitative properties are reduced to number, in the natural sciences properties are reduced to kinds. The analogy between systematic science and mathematical science is perfect, and both are partly conventional. As it is necessary to reduce properties to number in order to treat them mathematically, so it is necessary to reduce properties to kinds in order to treat them logically.

Bodies are composed of particles, and the elementary particles are probably alike. They have been reduced to about seventy kinds by chemical analysis. Logical analysis reduces them to one kind, and if it is valid then they are alike in being composed of one substance with like properties. If only the chemical analysis is valid, then there are seventy kinds, but they are alike in having the same properties, and unlike only in having different quantities or proportions of these properties. All bodies have a fundamental likeness in essentials, and a contingent unlikeness in relations. Every physical body is like every other physical body in its essentials and unlike in its relations.

The natural classes which exist and those which have existed in the past (for the processes of extirpation have always existed in the world) have a meaning for us in expressing the agencies which have been at work in producing the present stage of the world, for every gap represents some event of history. Planes of demarcation are thus landmarks of history to guide in research. As bodies have appeared and disappeared upon the stage of time and the actors changed with every act, a history of transcendent interest is involved, for in the discovery of classes we may restore the history of the earth.

It is seen that classification is the discovery of kinds in series. If classification is discovery, classes are not conventional but natural. In any stage of classification, while yet all of the attributes are not known, there may be imperfections in distinguishing kinds in series; the kinds depend upon properties, but all the properties may not be known, and there may be gaps in our knowledge of the series, so that imperfect knowledge is imperfect recognition of kinds in series; therefore, classification is always tentative by reason of imperfect knowledge.

When a classification is once established upon a logical basis, it need not undergo dissolution to be reclassified, for when the germs of classification are established on a logical basis it has but to grow with increasing knowledge.

While classification may grow it will always be recognized that there is but one system, as the individual is but one individual, though he may grow from infancy to maturity. The classification of which we speak is genetic, and while but one may exist that one may undergo changes on the way to perfection.

The test of classification is this: First, within the class all of the individuals must constitute an unbroken series, with a beginning and an ending, each class demarcated by a gap or discrete degree.

Second, the classes themselves must be seriated with the least possible gaps. Third, the series thus produced must be traced to convergence. A classification guided by these three laws is valid, when all the facts are known, and it is relatively valid when these laws are observed in the consideration of the known facts. The goal of the science of classification is to discover kinds in series and coördinate series of kinds in systems, and systems again in series.

In every perception there is a semblance of dichotomous classification of that of which the ego is aware, as distinguished from the environment. Such a process is involved in the first act of judgment, and continues to the end, but it is simply distinguishing the object of judgment from its environment or the world outside of the object. In perceiving the horse, the horse is distinguished from the rest of the environment, and in order that this may be expressed in speech some logicians speak of the horse and the non-horse, the tree and the non-tree, the house and the non-house. This is but a method of naming, but that which is expressed is the whole world except that which is included under the positive name. By this expression we must not conceive that the non-object in any way negates the object, nor that the object denies the existence of the non-object, but must consider the particle "non" as a device in naming. This method of naming is accomplished by another method in modern biological science when it speaks of the individual and the environment. In logic this method of naming has led to much confusion, and in the logic of Hegel it has led to strange absurdities, all of which are cleared away when the non-individual is called the environment.

This semblance of dichotomous classification has led to many errors, for the habit has been formed and philosophers have sometimes diverted the method from its use in perception and attempted a dichotomous classification of the universe. It has rarely been suggested as a complete system, but it has been practically used by many in this manner, and is still so used. Thus, we hear of space and matter as if space were not one of the properties of matter; we hear of motion and matter as if motion were not one of the properties of matter; we hear of time and matter as if time were not one of the properties of matter, and we hear of thought and matter as if thought were not one of the properties of animate matter. Would a sane person speak of the horse and head, the horse and body, the horse and legs, the horse and tail, and then consider the horse as one thing, the head, body, and tail as other things? Yet this is the error of those who consider matter as one thing and properties as other things. All such methods are not only vague and idle, but pernicious in that they deform all the concepts involved.

There is another method of dichotomous classification just as pernicious, exhibited in the attempt to classify the properties of matter as dynamic and static, which was Spencer's classification. Here forces and causations are classified in one group as dynamics, and kinds, forms, and thoughts as statics; thus the distinction between causations and force as categories are confounded, as also the distinction between kinds, forms, and thoughts. For some purposes of discussion a schematization may be of more or less value, but it easily degenerates into illogical classification, especially when it becomes the foundation of a philosophy. This classification is a relic from an earlier stage of philosophy when properties were confounded with qualities, and both properties and qualities were classified as primary and secondary, with sometimes a third class as secundo-primary.

There are only five properties, quantitative and categoric. As abstractions they are wholly unlike one another, but in the concrete they are identical, for every particle of matter and every body compounded of particles has number, space, motion, time, and, if it be an animate body, judgment. The properties, therefore, are phases of the same body, and their abstraction must be pentalogic. In the science of mathematics the four properties are always recognized by every physicist. During the latter half of the present century the fifth property has been clearly recognized in the new science of psychophysics, which seeks to measure mental operations and treat psychology mathematically. In this field of modern research a large body of literature is already developed.

Mill, in his work on Logic, groups phenomena in a dichotomous scheme as the simultaneous and the successive; this is not a logical classification of phenomena, but simply a device in naming. Other writers divide phenomena into the coexistent and sequent, using other terms for Mill's scheme, while Mill himself used it as a classification, and thereby fell into many errors of logic. Spencer used it also, but legitimately.

Names are developed before classes are logically

distinguished, and, although naming involves a mode of classification, many devices of naming are very illogical methods of classification, but still convenient in schematization; a schematic name, therefore, must always be distinguished from a classific name.

Often the term physical is used to distinguish certain properties from those which are called intellectual. This is not a logical classification of properties, but a convenient schematization which if understood as a classification leads to error. It always leads to error when the abstract property of judgment or conception is held to be a substance, and to exist apart from time, motion, space, and number, or from causation, force, form, and kind. Then thought becomes a ghost.

As classes are found in nature and discovered by science, so groups are also produced by art for a purpose. As the products of nature are used in art a regrouping may arise which has in view only the characteristics of the things of nature and art as they are utilized in art. The builder recognizes the group of building materials as a class of things in which he is especially interested; the mariner the group of stores which he must provide for his voyage; the traveler his outfit which he must carry in his trunk. Such groups can be illustrated to an indefinite extent. They are always dichotomous on the plan of perception which groups things into the perceived this and the not this, or the individual and the environment. The two groups are composed of heterogeneous things, as they are known in natural classification, selected for a purpose and distinguished from those not selected.

In the presentation of a theme the speaker or writer is prone to arrange his material in a scheme which may be very wise for the purpose intended for distinct presentation and clear understanding. Such a piece of valuable literature may live, and the schematization may be taken as a classification with disastrous results. Schematization is valuable for ephemeral purposes, but classification has enduring value. The author who uses a valid classification as a schematization is always clear, while the author who uses a schematization which is not a valid classification thereby introduces an element of confusion.

Before the rise of science artificial and natural classes were often confounded. This especially appears in the development of names. Among many tribes of Indians things are classified into the standing, sitting, and lying; or into standing, sitting, lying, and moving, which is a classification by attitudes. In other languages things are classified by their states. A fundamental classification existed among the Greeks as the four elements, earth, air, fire, and water.

As science first develops, classes are based on inadequate characters; that is, a few characters only are taken as the basis, as in the Linnean classification of plants. But as science progresses, classes are discovered which more thoroughly express the facts; to these classes names are given, and the names as they are thus classed are the names of the things classed and the metaphoric names of the concepts of the classes.

Now we must consider identity and difference. Mineral bodies are identical in having the four properties of number, space, motion, and time, and by hypothesis, judgment; but they differ in relations. An organic body undergoes a secular change in kind, form, force, causation, and by hypothesis, conception, and differs from itself at different times in these respects. At different times the same body in part is identical in its different phases and in part different; thus there is identity and difference in the individual at different times.

In the plant there is the same identity as in the mineral, but there is an additional difference, for the plant grows by minute increments through the addition of new matter.

The animal has the same identity and difference as the plant; but it has other differences, for the substance of the animal grows and decays coincidently. The same animal is not composed of the same identical substance from time to time, but only of the same kind of substance, for its food is continuously assimilated and used in function and discharged as new food is absorbed.

But there is another identity to be explained, namely, class identity, for the member of a class is identical with every other member of the class in some respects, and different from every other member of the class in other respects. In minerals the individuals are identical in being composed of the same substance, and different in being composed of different quantities of the same substance. The individuals of a class of plants are identical in substance, but different in quantity and in history. In animals the individuals are identical in kind of substance, different in quantity and history, and also different in that their substance undergoes a secular change by absorbing new substance and throwing off the old. In common ideation animals differ in other respects from plants and minerals, in that they are animate bodies, and have the property of judgment or consciousness.

The same body is relegated to different classes in a hierarchy of classes by the consideration of different degrees of identity. The fewer but more fundamental the identities the greater the number of the individuals in the class; the fewer the number of variables and the less fundamental the variables, the smaller the number of individuals within the class. Following the methods of classification as bodies are found in nature, the same object is found to fall within different classes, which constitute a hierarchy. Thus every object has its identities grouped in a hierarchy of classes. A horse is identical with all other horses in certain attributes, but it is also identical with all animals in a fewer number of attributes, though it may be considered as an object. No horse exists solely as an animal; but it may be considered only as an animal, that is, we may consider those properties which make it an animal. No horse exists which is only a vertebrate, but we may consider only those characteristics which make it a vertebrate. No horse exists only as a mammal, but we may consider only those characteristics which constitute the mammal. No horse exists only as a horse, but we may consider those characteristics which constitute the horse and still there will remain the characteristics which distinguish it from other horses. Thus, in the different groups into which the horse is thrown in the series, we may consider its different attributes in every class, but it is only a method of consideration. This is a concrete world, and objects are concrete in all their classes, and no entity or body exists which corresponds solely to the class to which the object belongs.

A fallacy has tainted philosophy from the early history of civilization to the present time through the entanglement which has arisen from considering an object as belonging to different classes. It has been supposed that there is an entity which represents the class as distinct from every individual of the class to which the characteristics of the individual adhere. This nothing which has been entertained by philosophers is a fallacy. It is an easy thing to be lost in the maze of speculation about classes in which fallacies fill the mind and obscure the real world. Abstraction is simply a method of consideration useful and necessary in cognition, but to suppose that the things which we consider abstractly have a disjunct existence is to enter the realm of metaphysical illusions.

In early society the origin of names was not understood, and often names were believed to be properties, especially when properties were considered as qualities. When the characteristics which belong to a kind and make it a kind were considered as the attributes of distinct entities, called essences, then the name was considered to be one of these essential attributes or properties by which the class was designated. Thus a fallacy was made to breed a fallacy, and the two fallacies grew up together and are often connected, and how can you dispel the fallacy of essence without dispelling the fallacy of inherent name? Thus a pair of ghosts stalk the world together, and fight each other's battles. How these ghosts waltzed in the dance of philosophy seems a marvelous feat—a Tam O'Shanter dance of warlock and witch.

It is not strange that those who believe in a substrate of substance should also believe in an essence of kind; then this essence becomes the noumenon, and the characteristics of class become the phenomena; this dream is the reality of metaphysic; the knowledge of science is the identification of phenomenon with noumenon.

It has already been asserted that classification is a tool of logic; and this assertion now requires demonstration. The first law of deduction may be formulated in the following terms: whatever is true of anything is true of its class identity. Inductive reasoning is the discovery of the members of a class; that is, it is classification; deductive reasoning is the application of the first law of reason as given above.

A drop of water is analyzed and found to be composed of oxygen and hydrogen in certain proportions; other analyses verify this conclusion. Now, by the first law of deduction every drop of pure water in the sea, on the land, and in the air has a like composition; but in every drop of water found in nature there are other substances, and for the analysis of the water these substances are eliminated. Now I take water from a spring, and though satisfied that water is oxygen and hydrogen in certain proportions, yet in this water there are other substances for which I must seek, and by induction I discover them. Induction is here the discovery of the nature of pure water and other kinds of water, and as these facts are learned by induction the several kinds are classified, and then the first law of deduction applies to each class. Induction is the discovery of class, and thus the discovery of the law; deduction is the application of law.

All laws may be reduced to this form, and are but variants of it. There is nothing occult or wonderful in the nature of law; law is just as simple as relation, just as simple as persistence, just as simple as speed, just as simple as extension, just as simple as unity. In scientific philosophy the process of reasoning reduces the complex to the simple. In metaphysical philosophy the attempt is made to explain the simple in terms of the complex.

Many errors have arisen in respect to the nature of classification, of which two are of such importance to our present work as to require elucidation. It has been held by some that classes are inventions and not discoveries, especially by those who have reified and personified the world as pure mind. Some who have not fallen into this error have still considered classes as artificial, invented for the purpose of economizing thought, and that real classes are found only because all of the units are not apprehended, and that classification is thus a product of ignorance and an infirmity of language. To a mind having infinite comprehension classification would be unnecessary; the whole would be grasped in mind simultaneously. Now ideas are evolved serially, hence it becomes necessary to take them one by one as they come and to group them and regroup them in hierarchies, for while the bodies of which they are ideas are presented to the mind serially of themselves, they exist in systems of hierarchies, and they are thus presented in nature in a hierarchy of bodies of different orders.

The things of this world are presented to the senses

in a chaos of phenomena. At every glance of waking life we see a number of heterogeneous colors and a number of heterogeneous bodies. While this goes on we hear a number of heterogeneous sounds arising from heterogeneous bodies. At the same time we smell heterogeneous odors from heterogeneous bodies, and taste heterogeneous flavors from heterogeneous bodies, and touch heterogeneous surfaces of heterogeneous bodies, and discover heterogeneous forces in heterogeneous bodies, perhaps all in one second of time; but as the instances come new sensations come in the most heterogeneous manner, and the things presented to the senses seem to constitute a chaos. Out of this chaos a cosmos arises, for sensation, which is the fundamental faculty of the mind, is classification. This classification is fundamentally mechanical. The eye sees the colors and classifies them, the ear hears the sounds and classifies them, the nose smells the odors and classifies them, the tongue tastes the flavors and classifies them, the touch feels the surfaces and classifies them, the muscular sense feels the forces and classifies them, and behold, all of these sensations are wrought into systems as if by magic!

In one chapter we considered bodies as particles, and found that we were discussing quantitative properties, as number, space, motion, time, and judgment. In another chapter we considered particles as incorporated, and found ourselves to be dealing with categoric properties, as kinds, forms, forces, causations, and concepts. Then in another chapter we discussed the reincorporation of bodies as they are revealed in geonomy, and found ourselves dealing with both quantitative and classific properties. In another chapter we discussed methods of reincorporation in plants, or the bodies of phytonomy, in which we were compelled again to consider quantitative and classific properties. Finally, a chapter was devoted to a third method of the reincorporation of bodies as they are revealed in zoönomy, and again we were led to consider both quantitative and classific properties.

Here it becomes necessary to more clearly distinguish those bodies which we have called molar, for the term has been used in a somewhat restricted sense which should be understood. By a molar body we mean one which is revealed to the senses without the use of instruments such as the telescope, the microscope, the spectroscope, or the crucible, aided by computation and logical ideation.

All geonomic bodies are molar bodies, and so are plants and animals. Savage and barbaric men supposed the stars to be molar bodies, while ethereal bodies were wholly unknown, their manifestations being interpreted as phenomena due to molar bodies. Thus the concepts of mankind were first compounded of judgments about molar bodies, or such as were supposed to be molar, and intellection progressed in this manner until the dawn of civilization and the invention of instruments of research, mathematical computation and logical ideation.

Man seems to occupy a position in the world midway between extremes of magnitude. On the one side there are bodies which are vast systems of stars like the solar system, and these are revealed by the employment of instruments as aids to vision, and are further revealed by careful investigation as magnitudes are measured and computed; on the other hand there are magnitudes that are so minute that they are revealed only by the microscope and other methods of investigation, especially in chemistry where molecules and atoms appear, and are further revealed when we investigate the nature of the ether and find ourselves immersed in the contemplation of magnitudes that are lost in immeasurable numbers. Between these extremes we find molar bodies that are revealed to the senses as bodies without the supplementary devices. Thus we use the terms molar, stellar, and molecular to designate in a general way the magnitude of bodies as they are compared with the magnitude of our bodies and the means by which these comparative magnitudes are determined.

When we go on to discover stellar bodies we find that we observe them from our standpoint by considering their quantitative properties, that is, considering them as particles, and ultimately find that these stellar particles are combined in systems. Again, when we investigate the minute constitution of bodies we also consider them as particles, and deal with quantitative properties, and through the quantitative properties discover their forms as structure and figure. Thus it is that in the minute and vast alike, in stars and in molecules, in systems and ethereal particles, science is interested chiefly in quantitative properties, and through them classific properties are revealed.

Plants and animals, which are molar bodies by our definition, first come to be investigated in modern or national civilization when they are treated as kinds and classified; but as we discover their kinds we discover relations of form, force, causation, and mentation, and a multitude of appliances for research are developed.

In these realms research deals with categoric properties, and reduces all phenomena to kinds, and the ultimate expression of all knowledge is classification verified by quantification. In plants bodies are reduced to particles when a minimum of computation can be used. So animals are reduced to particles by research, and again computation can be used. The goal reached by research is the particle, the way traveled is by classific logic, while in etheronomy and astronomy the goal reached is the body, and the road pursued is mathematical computation. In geonomy both methods of research are used. The quantitative and categoric methods of research are conventional. Quantities are measured by conventional or artificial methods, with artificial or conventional units. Kinds are also in the same sense and by equivalent processes selected as the representative of forms, forces, causations and mentations in order that classification may proceed and logical results be reached. Thus logic and mathematics are reciprocal methods of procedure in the cognition of the world. The mathematical method is chiefly deductive, the logical method is chiefly inductive, but they cannot be separated. There is no deduction without its reciprocal induction, and there is no induction without its reciprocal deduction. Deduction is abstraction which posits induction, and induction is abstraction when deduction is posited. Deduction and induction cannot be carried on apart, for deduction is dependent upon induction, and induction is dependent upon deduction, and the attempt to dissever them leads the mind into a fog of speculation where men are lost on the shoreless sea of metaphysics or the endless trail of unrelated facts.

CHAPTER X

HOMOLOGY

Extension may be defined as exclusive occupancy of space. The particles having extension exclude others from that extension, and thus extension has also been called impenetrability. The particle having motion, changes its position to occupy space vacated; hence, change of position is always exchange of position. As the particles are all in motion at an inconceivable rate of speed, one evacuates its position as another enters.

The idea of a plenum of substance was entertained by philosophers in the early history of civilization. Gradually this was abandoned by many, but lately it has been revived as best explaining the phenomena of the ether, and countenance is given to the hypothesis by the demonstration that molecular bodies have internal motions and interspatial ether.

Space is the relation of extension which particles bear to one another in position, when considered without regard to their incorporation in a higher body. If the particles be not ultimate a medium of smaller particles is intercalated. Space, therefore, is the extension of positions.

While space is the relation of positions, positions and relations must vanish if the extensions vanish. These relations may be relations of direction, or they may be relations of distance, but as particles are in motion the relations of direction are changed. In the same manner the relations of distance may change. Thus the boy and the dog may change relation of direction, when one or both move, and they may or may not change relations of distance at the same time. These space relations do not change by reason of intervening bodies. The boy may be a yard from the dog though a wall intervenes.

When positions are considered as established by incorporation, forms are observed having the relations of the particles established, and these established relations constitute structure and figure; thus form is figure and structure. When space becomes form, extension becomes figure and position becomes structure.

By incorporation particles retain in a qualified degree their space relations; that is, the space relations must be fixed within such limits that the incorporation is preserved, for if dissolution supervenes form relations are dissolved. Still, form relations are not fixed with such rigidity as to prevent internal motion. A body may still remain a body within certain degrees of temperature, passing through stages of bulk by contraction and expansion, but if the expansion is increased beyond the critical point the body is dissolved.

We consider bodies as particles when we consider their space relations, and we consider them as forms when we consider their corporeal relations as units. Habits of thought are formed in such a manner that some bodies are usually considered as particles, while other bodies are usually considered as bodies. By like habits of thought it is customary to consider the solar system, not as a body, but as an assemblage of orbs, for the science of astronomy has not yet successfully attacked the problem of the relation of the solar system to other stellar systems. When a body is considered as an individual in shape and structure, form is presented; but when a body is considered as a community of particles, space is considered. Thus it is seen that what is called space in the relations of particles, is called structure in the relations of form. In this treatise the term space is never used to denote the void—the nothing—but is always used to denote something real; so that space relations are the reciprocals of structure relations.

When we consider stars as such they are bodies, the particles of which are molecules. If we could study them as molecules they would present relations of structure; so we may conceive of such relations, though we cannot actually observe them; but we can observe the figures of the bodies. Stars are embodied into systems when they, in turn, become particles and have space relations to one another; this is structure from the standpoint of the system, but the systems as bodies have form as figure and structure. Here in the celestial realm is found a series or hierarchy of individuals and communities.

When we come to the study of the earth as a body, we find it composed of four particles: the atmosphere, the hydrosphere, the lithosphere, and the centrosphere. When we consider it as a body we consider form and structure; when we consider the spheres as particles their relations are those of space, one above another; thus in the body there is form, in the particles there is position, and that which is position in the particle constitutes structure in the body.

Again the stony crust or lithosphere may be considered as a body when its particles are formations of igneous, aqueous, aërial, vegetal, and animal origin. Then as a form its structure is derived from its formations, which are related to one another in structure.

The formations may be considered as bodies; then the blocks of which they are composed, called rocks, are particles. The structure of the formation is the arrangement of the rocks; the relations of the rocks to one another are relations of structure. We may consider rocks as bodies, and omitting ill-defined granulation and incomplete crystallization and also omitting for the present purposes the consideration of the substances of which the rocks are composed, we may consider rocks as bodies with particles of molecules; then the form of the rock is its structure of molecules; the relation of the molecules to one another in position is structure. Omitting various molecular stages in the hierarchy, we find atoms as the particles of molecules, the molecules having form in figure and structure and the atoms having space in their relations of positions to one another. Thus in the geonomic realm there is found a hierarchy of individuals and a hierarchy of communities.

The sciences of geonomy are divided usually into two correlative groups, called geography, in which five departments are pretty well recognized, namely, ethereal geography, stellar geography, aërial geography, hydrographic geography, and land geography; and geology, composed of five well recognized sciences: chemistry, mineralogy, dynamics, structural geology, and paleontology. What I have called geography is approached from the standpoint of quantitative properties, while those sciences which

I have called geology are approached from the standpoint of categoric properties. This division into two groups is well recognized when the one is considered as deductive and the other as inductive, or when the one is relegated to the physical division, the other to the natural history division.

We may consider a plant as a body; then the phytons of which it is composed are particles. A phyton may be considered as a body, then the cells are considered as particles; in turn, the cell may be considered as a body, then its blasts may be considered as particles. Then a blast as the nucleus may be the body whose particles are molecules, and the molecule as a body has atoms for its particles. Thus there is a hierarchy of bodies and of particles in the plant realm in which the bodies have form while the particles have space. We do not aspire to a treatise on botany, but stop to consider only certain facts which are essential to this argument; a consideration of the higher plants will serve our purpose. Certain phytons are modified to become roots, which are the organs devoted to the absorption from the earth of the materials which are to be woven into the plant; other phytons become the stem for support; others the branches for expansion; others the leaves for respiration; others pistils and stamens for reproduction, while others become floral envelopes for their protection. Every group of phytons in the plant, therefore, has a separate function, and is an organ. All of these organs, except those for reproduction, have functions relating to the metamorphosis of the individual; but the floral envelope and seed organs are devoted to reproduction. This development of phytons into organisms and organs leads in the study of botany to the consideration of the homologies of the organs. Reproduction in the plant makes a vast stride from ontogeny to phylogeny. Here we are introduced to the subject of heredity. Plants are multiplied in vast numbers and the offspring inherit likeness from parents; this inheritance is put at usury, so that each heir inherits the entire possessions of the legator, and wealth is multiplied by bequest. Then the legatee places his wealth at usury, and with its increments bequeaths it to every individual who is a legatee: so organs and organisms are developed.

The simplest plants are protophytes and unicellular; but these unicellular bodies are still more highly organized in the higher protophytes when unicellular bodies are connected with one another by vegetal threads which are themselves unicellular bodies metamorphosed by elongation, as in the slimes. The protophytes are simple cellate bodies which multiply by fission, and growth itself becomes reproduction.

The cells themselves are organized into tissues and the tissues are arranged in form as planes and combinations of planes. In combining, the planes are sometimes arranged about stems of trunks. These are the thallophytes. The entire thallophyte is a cell with structural parts as nucleus endoblast, mesoblast and exoblast.

In the thallophytes growth is chiefly marginal to a plane. Reproduction is not a division of the whole plant into new plants, but is a division of only portions of the plant which are organs of reproduction. Spores are thrown off from the surface of the reproductive organ.

Systematic botanists seem to be agreed in placing the bryophytes below the pterodophytes.

In the bryophytes a nucleated cylinder is produced which grows mainly by elongation. Special organs of reproduction appear with many devices for the preservation of the spores and their distribution over the soil. In the nature of these reproductive organs I find evidence of high rank. The leaves also are not mere fronds or expansions of the body, but are highly differentiated leaves.

In the pterodophytes the thallophytic structure in planes is still predominant, but roots are developed, the bodies are of more or less cylindrical form, and thallophytic leaves are often found as fronds. The reproductive organs are more highly differentiated. In some the margins of fronds are reflexed to make seed vessels, in others segments of fronds or entire fronds are transformed and there are other methods of forming seed vessels. In all a great variety of seed vessels are found, all exhibiting comparatively simple transformation; the cellate structure of the entire plant is still preserved, though greatly metamorphosed.

The spermatophytes are the flowering plants. In this sub-kingdom the seeds are no longer mere spores, but are plant bodies with microscopically developed forms. The entire plant preserves the cellate structure, while all the organs of the plant are of cellular structure.

The forms of plants are seriated three times:

First, there is the series through which the individual plant passes. Now the forms exhibited in the individual plant at different stages of growth may be compared with the forms of plants of the same species taken at different stages of growth, and the same results reached without waiting for the growth of one plant.

Second, we may study different species of plants and compare them with some one taken as a standard; but this should be a plant of the highest structure. Then in comparing plants of lower structure with it, it will be found that the stages marked in the growth of the higher plant are represented by stages in the order in which the record has been kept in the higher.

Third, a record has been kept in the tome of geology by which the forms of plants have been recorded, not in the language of symbols, but in the language of the forms themselves as fossils. While knowledge of this record is incomplete, in so far as it has been read, it agrees with the individual records and the class records.

The cell of the plant has a structure consisting of a threefold capsule or wall and a nucleus. The seed of the plant has the same structure with the threefold wall or epidermis and nucleus, and the cellular structure is preserved in the plant itself, which retains its envelope of bark divided into three layers which contain a nucleus. We have already found that the earth has a cellate structure, in the air, the sea, the land, and the nucleus; the elements of this structure we have called spheres or cellates. We call the structural elements of the cell, the seed and the plant, blasts or cellates.

Some plants are single celled. These have many forms, but one form is homologous with another, that is, it is composed of the same structural elements. The cells are compounded into phytons and grow into different forms, but one phyton is homologous with another; then phytons are compounded, and still higher plants are produced which are metamorphosed into different forms; but one higher plant is homologous with another. Phytons being composed of cells are homologous with cells, and higher plants being composed of phytons are homologous with phytons, and thus with cells; that is to say, the discovery of homologies in plants is the discovery of the morphologic elements of which they are compounded. As they are compounded, cells are differentiated, and when they are compounded into phytons differentiated cells make differentiated phytons, then differentiated phytons make differentiated higher plants.

In plants there is another set of homologies in the position of the leaves, which is revealed to us in the science of phyllotaxy.

Metamorphosis is growth and decay. One body cannot grow unless another body decays; one crystal cannot increase in size unless some other yields its particles for that purpose; one plant cannot grow unless molecules of water and other substances are used to constitute the molecule of protoplasm; one animal cannot grow unless some other animal or some plant dies; thus metamorphosis is decay of one and growth of another.

Development which supervenes upon metamorphosis is the production of coöperative organs all necessary to the life, growth and reproduction of the individual, and these organs have different powers, which in physiology are called functions. The exercise of functions is accomplished by metabolism, which is the recombination of chemical particles so that new particles come to take the place of those rejected. In this exchange particles do not lose speed, but all have their directions changed. That which is required for present consideration is that exercise stimulates the exchange. Now, activity of function increases metabolism; total rest from activity retards metabolism, and continued rest will ultimately cause atrophy; thus the form of the animal is transformed, for the slow changes that occur in this manner are transmitted to offspring, and if the offspring continue the process, growth or decay are continued in the next generation, and on through many generations, producing results as varieties and finally species, as organs are developed and extirpated.

We have now to consider animals and the organs of which they are composed in the transmutations through which they pass as illustrating the subject of morphology.

There are five great classes of animals: Protozoa, Radiata, Mollusca, Articulata, and Vertebrata. The Protozoans are unicellular or simple combinations of cells. Above the Protozoa, animals are organized on four different plans of structure, but they are all compounded of cells, though many of the cells are greatly modified. In these modifications the cellate structure reappears as a fundamental homologue in every organ of all of the higher animals, and it is still found in the animals themselves. The phytons of plants are the homologues of organs in animals. There may be many phytons serving the same functions in plants, as there may be many organs serving the same function in animals; but in animals, as functions are differentiated, kinds of organs are multiplied and the number of organs performing the same functions is diminished from the lower to the higher organism.

In animals the fundamental homologies are found when we discover that all organs are dermal. We cannot stop here to make an exposition of this subject throughout the whole animal kingdom, but will confine ourselves to one small group of vertebrates, namely, the mammals.

First, there are organs of nutrition, constituting all those that take part in the digestion, secretion, and excretion of food. Second, organs of circulation, by which the food when prepared for assimilation is distributed to the tissues. Third, organs of locomotion, constituting the muscular, tendonous, and osseous systems. Fourth, the reproductive organs. Fifth, the organs of mentation, constituting the nervous system.

The organs of digestion which prepare the food are severally sacs and tubes, and conjointly they constitute a system of sacs and tubes, but in this system locomotion must be accomplished, and hence a muscular system is attached to the digestive system. Thus all the organs of digestion are cellate in that they have the cellate elements, for they are composed of encapsulated parts, or inclosing or inclosed envelopes.

The circulating organs are all found to be cellate as tubes or sacs, one or both. In this system extreme variations are found; in the veins and arteries the tubular structure is carried to its highest development, while in the gall, the liver, and the lungs, the sacate form is observed; while the heart is a muscular organ it is still provided with tubes and sacs.

In the muscular system every distinct muscle has a cellate structure, and they are compounded into groups on the cellate plan. Muscles when considered in phylogeny are found to develop into tendons and tendons into bones; the same development is discovered to a limited degree in ontogeny, so that muscles, tendons and bones are homologous. The cellate structure of bones is conspicuous, for they all have the periosteum and nucleus.

In the reproductive systems both sacs and tubes are found, all of cellate structure.

In the nervous system the differentiation between sacs and tubes is carried to its highest degree. The nerves proper are all tubular cellates. In the lowest units they are cellate, and they are compounded as cellates. In the ganglia they are sacate, and are compounded as sacs. Certain of the ganglia have osseous protection as vertebræ, and every vertebra is a cellate structure as a bone with elaborate differentiation in morphology. The vertebræ that have united to form the cranium are extremely differentiated as morphologic elements, but the most extreme of morphologic elements is found in the organs of sense, every organ having a distinct form, and all preserving the cellate structure.

Then the systems of organs which we have just described are themselves compounded into systems, of which hint has already been given. While this subject is vast and tempting, the purpose is subserved merely by giving a few illustrations; and we must forego systematic treatment. In the mouth there are found elements of the digestive apparatus: the circulatory apparatus, the muscular apparatus, as muscles, tendons, and bones, and perhaps elements for reproductive purposes and certainly apparatus for mental functions in the sense of taste. Perhaps in all parts of the body all the five functions are performed by apparatus provided for the purpose. Finally, the entire animal has a sacate and tubular structure, and is thus a grand cellate of a high order of compounding.

The cellate homologies of the man are repeated in all mammals, while the same facts can be seen in birds, reptiles, batrachians and fishes, for all the pentalogic classes present a vast hierarchy of homologies, which illustrate the theme of morphology. Nor does the subject end with vertebrate morphology, for the theme is illustrated in the homologies found through articulates, mollusks, radiates, and protozoa. That which we find in the pentalogic classes of plants we find also in the pentalogic classes of animals—a vast hierarchy of homologies.

Perhaps the great field yet to be cultivated in morphology is in the study of the articulates, especially among insects. The sudden transformations which they undergo in their life history permit the examination of morphologic stages to such an extent that morphology can be studied with all its multitudinous phenomena, and a wealth of science has already been accumulated as a heritage for the army of scientists necessary to give us a complete account of the insects of the world, among whom are found tribes that vie almost with men in demotic development.

We now see how homologies are extended from atom to organism. There are homologies discovered in the atoms, which has given rise to the theory that the atoms discovered in the seventy substances are not ultimate particles, and it must be remembered that it rests only upon the validity of reasoning from homologies, but that all deductive reasoning is based on homologies; it may, therefore, be impossible to reach an inductive demonstration of the complete homology of ultimate particles, but the deductive reasoning is perfect. Then molecules which cannot be seen and cannot be manipulated as individual, but can be discovered only by chemical apparatus, are found by analysis and synthesis to exhibit many homologies, and the science of chemistry undertakes this enterprise.

The earth is a cellate body, and from facts revealed by astronomy it is confidently affirmed that the stars are cellate bodies. Finally, homologies are found in plants and animals; thus there is a hierarchy of homologies throughout the universe which constitute a continuum, and logically no plane of demarcation can be discovered which constitutes an absolute gap. The continuum is not completely demonstrated by induction, but is abundantly demonstrated by deduction.

Homologies have a high development in the organization of demotic bodies discovered in the animals, especially as they are represented among the higher insects, but more fully illustrated in the organization of human society. The forms of organization are various. In the tribes of the world families are organized into clans, and clans into phratries, and phratries into tribes, and tribes into confederacies. In passing from savagery to barbarism, the clan becomes the gens. In all the multitudinous forms of tribal society, homologies have been discovered. In the family husbands and wives, parents and children are found, and sometimes grandparents and more remote kindred are included. In the gens consanguineal kinship is reckoned in the female line; in the tribe it is reckoned in both male and female lines, and ties of affinity are observed. In the confederacy conventional kinship is recognized, and other homologies exist in multitudinous ways. For example, relative age is recognized in the family, in the clan or in the gens, in the tribe and in the confederacy, and to carry out the homology age is often determined by convention.

In national organizations another set of homologies are founded on those of tribal organization. Thus, in the United States we have the family, the township, the county, the state, and the nationality, and homologous units are found in all civilized governments.

Whenever two or more bodies are homologous they are identical, though they may at the same time be different. Homology in form is thus the reciprocal of likeness in kind, so that homologies fall under the same law with kind, and it may be affirmed that whatever is true of an object is true of its homologue in so far as they are identical, which is but another statement of the law already given in classification, that whatever is true of a thing is true of its class identity. We have seen that there is a vast system of homologies extending throughout the universe, commencing with perfect homology in the simple element; but gradually differences appear, becoming more marked as compounding proceeds and differentiation is more marked, that is, there is successive progress in variation from the simple to the compound, and this variation appears as increasing complexity. As things become compound they also become complex.

In the foregoing chapters an attempt has been made to show the relation which exists between extension and unity, position and plurality, space and number, form and kind, together with metamorphosis and metalogisis. Now it remains to show the relation between organism and class, together with a general statement of the relation between morphology and classification. It has been shown that a class is a series of kinds, and as a series it is a disjunct group in a more extended series. It has also been shown that a form undergoes a metamorphosis, and that an organism in its history represents a hierarchy of metamorphisms as exhibited in homology. Now, we must observe that through morphology classes are multiplied, for not only are kinds and series classified, but forms are also systematically grouped.

To investigate the structure of plants we dissect them, and find that when the limit of cell structure is reached and molecular structure appears, we are compelled to pass from dissection to chemical analysis. The highest molecule is protoplasm, but the protoplasmic molecule is composed of molecules of still lower orders until atoms are reached, when chemical analysis fails and only logical analysis seems possible.

In investigating the homologies of plants and plant structure we are thrown back upon the discovery of likeness and unlikeness, or, in other terms, of identity and difference; and we reason about plants, as these identities and differences have been discovered. The discovery of these identities and differences is induction, the application of the laws discovered is deduction.

What, then, is the significance of all these facts, and why should we gather them from the highways of morphology but for the lesson which they teach, that all forms of animals, plants, rocks, and stars are traced to the substrate of extension in the particle? Extension traced through all its complicated relations of space, form, metamorphosis and organism is found to be the ultimate substrate of them all.

Many extended particles incorporated in many bodies have relations of position, space, form, metamorphosis and organization, all of which are included under the term morphology. These relations cannot exist by themselves, but can only be considered by themselves, for relations of morphology are concomitant with relations of classification, dynamics and evolution in the concrete world. Bodies can be analyzed only into particles, and the particles still retain their properties, which may be considered abstractly. If I were called upon to nominate the fundamental error in the logic of transcendental philosophy I should name it the failure to recognize the distinction between analysis and abstraction. The failure to see this distinction seems to have led Pythagoras to found a philosophy upon number; it surely led Plato to found a philosophy on form; it seems to have led Aristotle to found a philosophy on force, and without doubt Spencer fell into this error; while it led the Scholastics to found a philosophy upon being, and finally it led the Idealists to found a philosophy upon thought. Thus the five properties of matter have every one in turn been taken as the substrate of a philosophy, and as the substrate was an abstract the philosophies have been abstractions. Metaphysics has been the attempt to found a philosophy upon an abstract unit, but science is the attempt to found a philosophy upon a concrete unit.

In this chapter an attempt has been made to make a summary exposition of the science of morphology, for the purpose of showing the certitudes which inhere in the science as distinguished from the illusions of mythology defended by speculative philosophy. In transcendental metaphysics the realities of the world are held to be phenomena in the sense that they are illusions, and are distinguished from noumena, which are the realities. Science deals with phenomena, and scientific men hold that phenomena are realities and noumena in the sense of occult substrates are illusions. Transcendental philosophy deals with noumena, and holds them to be realities, and deems phenomena to be illusions.

This is the issue between science and speculation, and the contest is war to the knife of logic against war to the blade of dialectic; but the knife has form, while the blade has void.

In science one noumenon is space, the reciprocal of form; the corresponding noumenon in metaphysic is space as void. Void space is a natural fallacy to men in savagery, while yet the presence of the ambient atmosphere is unknown, and the surface of the earth seems to be an empty theater for breath, wind, and storm existing as disparate bodies having a ghostlike existence. Having imagined an empty space, it still continues to exist in mythology as a void for the theater of gravity, heat, light, electricity and magnetism, after the air itself has been discovered and understood by all civilized men. Now that this notion is dispelled there is no void within the ken of man. All known interspaces have been resolved into forms. If in the depths of the infinitesimal void spaces exist between the particles of ether, it may be well to await their discovery ere we characterize them by assigning properties to nothing.

CHAPTER XI

DYNAMICS

A citizen of a township must obey the laws of the township. The same person is also a citizen of the county subject to the laws of the county, a citizen of the state subject to the laws of the state, a citizen of the United States subject to the laws of the United States, and finally he is a citizen of the world, subject to international law. Thus a man belongs to a hierarchy of governmental incorporations in which he may demand rights and must perform duties of allegiance.

In the same manner every atom of matter in the lowest body exists in a hierarchy of bodies. An atom of hydrogen exists in the molecule of water. The same atom exists also in the sea, the earth-moon body, the solar system, and the galaxy. Now this atom of hydrogen partakes in the specific or special mode of motion of every body in this hierarchy. We may consider the motion of the atom of hydrogen in the atom itself, if it is a compound body as some chemists suppose; then we may consider it in the molecule, then in the tide, then in the earth in rotation, then in the earth-moon body on an axis within the earth, then in the earth in revolution in the solar system, and then in the galaxy with the solar system, and if there be a system of galaxies we may consider it in such body.

This atom has components of path in an atom, in a molecule, in the tide, in the earth, in the earth-moon body, in the solar system body, in the galaxy body, and finally in another system which includes the galaxy, if there be such a system. If we consider the path of an atom in any one of the incorporations in the hierarchy, we can describe it in terms of dimensions of space, as space is limited by the periphery of that particular body; but when we attempt to describe its motion in two different members of the hierarchy, we are compelled to enlarge our conception of space, for the path of a particle in the atom is modified by its path in the molecule. Then if we consider the path of the atom in the tide we must still further modify our concept of it; then if we consider also the path of the atom in the terrestrial motion about the axis of the earth, we must again modify our concept of it; then if we consider also its path in the earth-moon body, the solar system body, and the galaxy body, we have at last a concept of the path of the atom in a hierarchy of bodies. If we desire, therefore, to conceive of the path followed by the atom of hydrogen directed by all its incorporations combined, we must imagine it determined by all bodies of the hierarchy, and thus to be spiral or vortical. I shall hereafter call this path a hierarchal path.

Descartes conceived this path to be vortical, and taught that the ether in moving in a vortical path carried with it the celestial bodies, and thus explained their revolution. I believe that he properly conceived the nature of the path which a particle describes in a hierarchy of bodies, but of the cause of this path he was in error when he considered that the whole body of ether describes the same path in a vortex.

We may describe the motion of a particle in any one of its incorporations, neglecting it in the other members of the hierarchy, and such a description is legitimate if it be understood as motion in the one incorporation; or we may describe the motion of a particle in two incorporations, but in order to do so it is necessary to use the terms of the space of the higher incorporation. This plan must be continued through all the incorporations if we try to describe all of the deflections of path which are experienced by the atom. If we consider the path of a particle of matter in every one of the hierarchy of bodies severally, we get as many systems of motion as there are bodies, and they seem, when thus narrowly and imperfectly considered, to be incongruous; but when we consider all of these paths concomitantly as hierarchal motion in terms of the space of the highest body, they are made congruous.

Every particle in the universe is in motion, which motion is probably constant in rate of speed. Motion is not only speed, but also path. While the speed in the ultimate particle is constant, the path is variable in direction. This is the proposition I am trying to maintain.

Of ponderable matter, as it is found in terrestrial and celestial systems, all particles are making a grand excursion of the universe. There is no star that does not proceed on this journey, nor is there any body of matter in the earth which does not proceed with the earth in its journey. Ethereal matter does not seem to proceed in this manner from position to position throughout the universe, but the motion of each particle seems to be confined to an environment of other particles, and vibrates back and forth or around and around within its narrow environment. A particle of ponderable matter never returns to the position which it occupies at any one instant of time, so far as we can determine by reasoning. Every position occupied by a particle is instantaneously evacuated, and another particle, either of ponderable matter or of ethereal matter, takes its place.

As there is a hierarchy of bodies, and as there is a hierarchy of paths for every particle of ponderable matter, so there is a hierarchy of freedoms of motion. Take three rods, fasten them together by their central points so that they extend in coördinate directions. The three rods will constitute a body of rods, and although the three are incorporated, that is, fixed to one another, the body has three degrees of freedom. Fix the ends of these rods to a stone quarry, and the three-rod body becomes a component part of the earth body, but still has three degrees of freedom. Then the same three-rod body has three degrees of freedom in the earth-moon body, the solar system body, and the galaxy body. Now we are compelled to believe, by reasoning based on facts observed in modern time, that the molecular bodies and the atomic bodies of the three-rod body have every one three degrees of freedom. This reasoning in molecular science is no less cogent than that in astronomical science, for chemistry gives the same freedom to atoms and molecules that astronomy gives to stars and systems.

We are compelled to conceive of the rigidity of the solid state as the homologue of the astronomical state, and as we know that the rigidity of the astronomical state is a mode of established motion, so we conclude that the rigidity of the solid state is a mode of established motion. Thus the concept is made that man stands between two realms of bodies, the vast or astronomical and the minute or molecular, and that which is observed in astronomy is repeated in chemistry. The astronomic world is the correlative of the molecular world. If there is no gap in this reasoning every particle of matter has a constant rate of speed which is subdivided among the paths of the hierarchy of incorporations to which it belongs. To this form we are compelled to reduce the concept of the persistence of motion or the correlation of forces; for if speed is constant in the atom the forces of the universe are correlative, or, to use a better term, are reciprocal. This conclusion that speed is constant in the particle is necessitated, and hence is valid if we accept the fundamental doctrine of modern chemistry that bodies are composed of discrete particles.

Motion can be diverted in any body of the hierarchy without increasing the speed of the particle. Nature never seems to add to or to substract from the speed of the particle, although the motion of a molar body may seem to be derived from another body so long as we consider only the molar motion. But when we consider the motion of the particles of the body in their higher and lower incorporations, we find that the apparent added motion is deflection. This is illustrated in the earth-moon body when it rotates about its axis, and thus deflects the motions both of the earth and the moon in their common paths around the sun. So, if a body suspended above the earth falls to the earth, its path with the earth in its course is deflected, and the path of the earth in its course is also deflected. In a falling body we observe not only the deflection of terrestrial motion, but the falling body itself is composed of molecules and atoms which are in motion, and the earth also is composed of molecules and atoms in motion, and these paths are also deflected by the falling of the body. The deflection of their terrestrial motion is but the reciprocal of their deflection in molecular motion. When a body, say of water, loses heat it gains the strength of structure, which is a force, and hence a mode of motion which it exhibits as ice. The body does not transmit its speed of particle to another body, but only induces a corresponding change in that other body from solid strength or rigidity to heat motion by deflecting molecular paths. Thus motion as speed cannot be dissipated. When water is evaporated the particles of vapor which are produced still have the same amount of motion as speed, and when water and carbonic acid are built into wood, their motion remains as the solid strength of the wood in another mode of molecular path. Here we see that rigidity or solid strength is a mode of motion as path. Thus it is that motion as speed is persistent in the particle, but as path it is variable.

Every particle in the wooden ball rolling on the floor has astronomical path, molecular path, and molar path. Consider one of these particles moving with the three kinds of motion as three constituents of path, and we realize that its speed is very great, and that the path which it traverses is greatly composite; that is, composed of deflected parts, in a hierarchy of bodies. If such a particle had its composite path straightened into a right-line path it would quickly pass out of the sphere of the solar system from whatever point within the system it might start, and in whatever direction the right-line path extended. But the molecule remains within the solar system because its stellar path is composite, and it remains within the ball because its molar path is composite, and it remains within the molecule because its molecular path is composite.

When the ball was started molar path was developed, and when it stopped that molar path was ended. We must not suppose that molar motion as speed came out of nothing and vanished into nothing. We resort to preëxisting molecular motion to explain it. We say that the molar motion was derived from the molecular motion of the hand that set the ball rolling, and that it was transformed into molecular motion in the wall which destroyed the molar motion. In making this explanation we assume that motion as speed went out of the hand into the ball, and then out of the ball into the wall. Is this true? Was the speed of the molecular motion in the hand diminished and the speed of the molecular motion in the wall increased? Did motion as speed go out of the hand into the ball? There was a change in the motion of the hand, and a change in the motion of the ball. In what did this change consist? We know that in part at least it consisted in a change of paths. The molecular paths in the hand must have had their directions changed, and the molecular paths in the ball must have had their directions changed. Is this change of direction all, or is there a transference of speed so that one loses while the other gains? The whole problem is narrowed to this issue: That which we call acceleration pertains wholly to deflection, or in very small part to speed, as loss of speed by one and gain by another.

There is still another set of relations to be considered. A body is composed of particles; in order that they should remain within the sphere of the body their paths must be composite, and in order that their paths may be composite there must be a sufficient number of internal collisions to deflect them and retain them within that sphere. If the body itself is moved the paths of the several particles in the average must thus be rendered less composite; that is, the number of collisions must be diminished. The motion of the body as such, therefore, is accomplished by diminishing the deflections within the body, and thus straightening their paths. The translatory motion of a body is a straightening of the paths of the particles of which the body is composed.

Imagine a man walking in a circle of ten feet radius. The sphere of his motion is within the circumference. He may soon walk a mile and never be more than twenty feet away from any given point in the circumference; change his direction so that his path is straightened, and he may soon be a mile away. A body of men walking in a circle remain together as a body within the circumference of the circle as it moves with the earth; change their paths to a cycloid directed to a distant point, and the body of men will move away in that direction; change their paths to parallel right lines, and as a body they may soon be a mile away and still in a circle. A division of an army may be maneuvering in a field as divisions, brigades, regiments, battalions, companies, and platoons, and yet remain in the same field enclosed by a wall; without walking the individual men with any greater speed you may march them to another twenty miles away, and they will lie down to rest at night with no less fatigue than if they had been maneuvering in the enclosed field.

In the same manner the molecules of the wooden ball are in motion within the theater of the ball, so that they do not pass beyond its boundaries; yet impose upon each molecule a change of direction in such a manner that they all move a little more in one course, and a translation of the ball is affected by a change of direction in the motion of its constituent molecules, and the ball still remains as an incorporated body. It is thus possible to explain the molar motion of the ball as a change in direction of the motion of its molecular parts, without assuming an increase of speed in the parts, but only a development of speed in the body by the deflection of its particles. By such an assumption the molar motion perceived by vision would be legitimately derived from the molecular motion known by higher reason, and appear as a change of direction in the molecular motions of the ball. No motion as speed would be created or destroyed, while the apparent molar motion would be explained by a change of direction in molecular motions, very minute as compared with the composite paths of the several molecules and atoms.

When we consider the total motions of the atoms of the ball shot from a cannon's mouth, an inconceivably small change of direction in the motion of every atom, as compared with the complexity of its path, would fully account for the flight of the ball as projected by dynamite.

Now we know of deflection and that it arises from collision, and we know of no other change in motion. Acceleration as increase of speed cannot, in the nature of the case, be demonstrated, for it may always be explained as deflection, and can never be explained without deflection. If acceleration is explained as deflection, it is explained by referring it to a known cause, and adequately explained.

It is illegitimate to assume an unknown and unknowable cause when a known cause is sufficient for the explanation. We may, therefore, affirm that the acceleration of a body is the deflection of its particles.

At the Brooklyn meeting of the American Association for the Advancement of Science in 1894, I read a paper on this subject, in which I tried to demonstrate that motion is constant in the particle. In the foregoing statement I have put this demonstration in another form. I now propose to give it in a new form by the method of *reductio ad absurdum*.

Newton taught that inertia is resistance to change of state, either as rest or direction of motion, and Newton also referred to the ambiguity of the term rest without pointing out the nature of this ambiguity. We have seen from the foregoing discussion that rest is absence of molar motion, and that molar motion is created by deflecting molecular motion. Hence the acceleration of a body is reduced to the deflection of its particles, as we have already seen. Following Newton, it is taught in the text-books of physics that inertia is resistance to deflection and acceleration; therefore, reduced to the simplest terms, inertia is resistance to deflection.

PROPOSITION

When two bodies collide their particle paths are deflected, but their particle speeds are unchanged.

First, assume that one body, A, has the mode of motion called rest, and that after the collision it has molar motion; then its molecular motions are deflected. Then assume that their speeds are accelerated; then the particle motions of B also must be deflected and accelerated, if action and reaction are equal in deflection and speed. Therefore, motion as force is created, which is absurd. But Newton's law says that action and reaction are equal and *in opposite directions;* therefore, action and reaction result only in particle deflection.

Second, assume that A is at rest, and that at collision B is brought to rest, and thus that B has the speed of its particles diminished; then motion as force is annihilated, which is absurd, but action and reaction being equal as deflection no speed is lost to either.

Third, assume that the particles of A are deflected and their speed accelerated, and that the increase of particle speed in A is derived from the particle speed of B; then action and reaction as speed are not equal, but while both are equally deflected A has more speed, B less, and the more equals the less, with opposite signs. Then A after collision, having more speed than B after collision, has more inertia, which is absurd; *therefore, when bodies collide their particle paths are deflected, but their particle speeds are unchanged.*

Let this argument be stated in brief:

First, the tendency of modern investigation is to explain all forces as derived from modes of motion. Great progress has been made in this direction, and the theory is widely accepted.

Second, all understood forces are collisions.

Third, if all forces are collisions the motions from which they result obey the third law of motion, that action and reaction are equal and in opposite directions. By this law it is seen that no motion as speed can be lost or gained by any particle of matter.

Fourth, by collision paths can be changed, but motion as speed cannot be transmitted by one particle to another.

Fifth, in starting or stopping molar motion there is an apparent creation and annihilation of motion, but this appearance is known to be an illusion. It is known to be in part deflection, and can all be thus explained; and if the third law of motion is valid it is thus explained.

It must clearly be understood that the above argument does not deny that molar motion as speed can be created or destroyed; it simply affirms that molar motion cannot be created from nothing, and that it cannot be annihilated, but that it comes from molecular motion and returns to molecular motions. Every particle of which we have knowledge is a constituent of many bodies in a hierarchy of bodies, and what is here affirmed is that the acceleration of a body in speed is deflection of its particles, that the particles themselves are not accelerated in speed, and further that embodiment itself is always a result of deflection in the particle embodied. A molar body may have its molar motion increased or diminished in speed by deflecting its molecular motions. If the speed of a molar body be changed, the direction of its molecular particles must necessarily be changed. This proposition is self-evident. The third law of motion is equally simple. The law here demonstrated affirms that acceleration in one embodiment is deflection in another, and it makes valid Newton's law, which would be an absurdity were the law here demonstrated untrue; and if untrue, the persistence of motion is an absurdity, and with it the persistence of energy falls to the ground.

When the concept of persistence of speed in the particle is once gained, there follows from it a series of corollaries which are demonstrations of axioms of scientific experience, but which otherwise have no demonstration. The following are examples:

PROPOSITION

Gravity, as inversely proportional to the square of the distance, is persistent in the mass.

Assuming that force is motion and gravity force, then if the particle can lose any of its speed it can lose gravity, which is absurd; and if in the collision of a body speed is transferred from its particles to the particles of another body, then the other body must weigh more, which also is absurd; therefore, *gravity, as inversely proportional to the square of the distance, is persistent in the mass.*

Speed is not a property which can run away by leaping from one particle to another and from one body to another; it is not an occult something—a mystery, a nothing. It is the speed of a particle.

We have seen that when particles in motion have incident paths they collide and their paths are deflected; hence, all motion is directed motion. Collision or impulse is the first mode of force in which action and reaction are exhibited. Then we note how right-line paths are divided into components by collision, becoming deflected paths; then how by systematic collisions they may be developed into revolution. Then we consider that particles may be incorporated in a body with their several particles revolving around a common center, and this revolution of the particles is rotation of the body. Thus by incorporation the motions of particles may be correlated by rotation and revolution, as exhibited in celestial bodies.

In the case of two stellar orbs revolving about a common center, as the earth and the moon, it is plain that gravity causes the deflection of both bodies inversely proportional to their masses. Here acceleration is chiefly deflection, being positive at perigee and negative at apogee. So, in the revolution of the sun and the earth about a common axis, acceleration is chiefly deflection, being positive at perihelion and negative at aphelion. Thus we have a well-known astronomical example of acceleration, and find it deflection and increase or decrease of bodily speed, and now we must refer this acceleration of speed in the body to deflection in the particles of which it is composed.

It is taught in astronomy that in the revolution of a planet the area of the radius vector is equal for equal times. This doctrine is made simple and plain when the nature of acceleration is understood.

In an ethereal medium of particles moving with a persistent speed, two bodies will mutually intercept collisions with the ethereal medium inversely proportional to the square of their distance apart, which is an explanation of the law of gravity, and is the theory of La Sage in terms of motion.

On page 642, Vol. IV, article 22, of Bowditch's translation of La Place's *Méchanique Céleste* it is stated:

"If gravitation be produced by the impulse of a fluid directed towards the center of the attracting body, the preceding analysis, relative to the impulse of the solar light, will give the secular equation depending on the successive transmission of the attractive force."

After proving this proposition and obtaining the secular equation of the attracting body from the successive transmission of gravity, the cause of the moon is discussed, and La Place decides that:

"We must suppose that the gravitating fluid has a velocity which is at least a hundred millions of times greater than that of light; or at least we must suppose, in its action on the moon, that it has at least that velocity to counteract her gravity towards the earth. Therefore, mathematicians may suppose, as they have heretofore done, that the velocity of the gravitating fluid is infinite."

The theory of La Sage is stated in terms of a fluid transmitted from one body to another. We now know that waves, not fluids, are transmitted in the case of heat and light, and in a like manner gravity as deflection must be considered as wave action or vibration in some form. With these principles the instantaneous action of gravity is simple and self-evident, for speed is not transmitted, but only deflection is caused.

Every particle has constant motion as speed which cannot be increased or diminished, and the absurdity of perpetual motion should be called the absurdity of perpetual collision between two bodies without other deflection. The particles collide because of impinging paths; they are deflected and their paths are turned apart, and they cannot be made to collide again until other external collisions bring their paths together. If the particle A is deflected after one collision, to be once more deflected, another collision is necessary. It is thus that the absurdity of perpetual collision can be simply demonstrated.

After such an analysis the doctrine of virtual velocities is self-evident; and there are many other consequences of this law which, properly understood, would make many propositions of physics self-evident.

Motion as speed is constant in the particle. The particle, of whatever order it may be in the members of the hierarchy, is accelerated by deflecting its particles. The principles or laws of dynamics are all corollaries of this fundamental law; hence dynamics may be taught as a deductive science. Thus we have the mathematics of number, the mathematics of space, and the mathematics of motion, all fundamentally deductive sciences.

CHAPTER XII

COÖPERATION

We have already discovered the nature of motion in its absolute as speed and its relative as path. The speed of the ultimate particle has never been measured; but bodies as such have their specific speeds and one is greater than another. Speed of a body is the rate at which it changes its position, regardless of the change of position of its particles to one another. The speed of one body may be taken as the measure of the rate of speed of another, and the process used gives rise to the formula of $L \div T$. The length of path is divided by the time in which it is traversed. Thus to convert motion into number it must first be converted into terms of space.

We have discovered, in preceding chapters, the transmutations which motions undergo by incorporation when they become forces. In order that they may be treated mathematically, it is necessary that they should be resolved into the quantitative categories and expressed in numbers. This resolution is accomplished by measurement, and different formulæ are employed which in mathematical science are called the equations of acceleration, force, impulse, energy and power. They are all devices for reducing force to motion and motion to number.

In molecular bodies motions are correlated in a manner yet unknown, but molecules are known to have interior motions exhibited in response to motions in the ether as its particles impinge on ponderable matter. The correlated structural motions of the molecule may be transmuted by collision with ethereal particles and be converted into heat—a mode of motion—so that which is structural motion will appear as heat, and if the transmutation is carried to a sufficient degree the structure of a molecular body will be destroyed, for by heat molecules are reduced to lower molecules or to atoms. Thus what appears in the molecule as structural motion appears in the particle as heat; and when disparate particles are incorporated in a molecule heat becomes molecular or structural motion. This may be stated in another way. By incorporation vibratory motion becomes structural motion; by decorporation structural motion becomes vibratory motion. We know that in stellar systems that which is structural motion in the system is vibratory or rhythmic motion in the particle; and we may conceive that stellar rhythms might be so modified in elongation or other ways that the structure of the system would be destroyed. Hence we may conjecture that in the molecule the rhythms of the particles become the structure of the molecule when these rhythms are systematic. There is much in the phenomena of motion which suggests that such is the case. In a previous chapter a brief statement was made to exhibit the universality of rhythm. That structural motion is always systematic vibration seems worthy of acceptation as a working hypothesis.

The form of force known as energy may appear in another phase as a succession of distinct forces impinging upon a single body producing effects which remain with that body. Energy in this phase is called process; thus a succession of waves of air may beat upon a tree and then action and reaction are successively involved in vibration. It is a process by which gravity deflects the stars into revolutions and it must always be a process by which particles are deflected while they are incorporated in bodies. A multitude of processes appearing in inorganic nature have already been exhibited, while processes which appear in the vegetal realm were noted.

In nature processes are developed into modes of force known as powers. The meteor falls upon the earth and acts as a hammer. Boulders are carried by streams and act as hammers and produce effects as such which the particles acting separately could not produce. Thus collisions which might result simply in deflection if the particles acted severally, produce fracture when they act conjointly. Particles may produce pressure when they act separately, but when they act conjointly pressures may lead to rupture. By the device of the lever forces are multiplied in effect without increase or diminution of force as such; the same is true in the pulley, the wedge and the screw.

All directed motions are motions subjected to conditions. These conditions are causes which produce effects, so that the consequent condition differs from the antecedent condition; that is, the effect differs from the cause. Two bodies collide and their paths are deflected; the antecedent direction differs from the consequent direction. Thus forces are motions subjected to causes which produce changes of condition which we call effects. Here we see again that there can be no motion without causation, and while they cannot exist apart, they can be considered separately; but the separation is only ideal.

It is now proposed to give an outline of the forces as they appear in the different realms of nature to exhibit the universality of coöperation.

In the ethereal realm we recognize light, magnetism, heat, gravity and electricity. These are usually known as motions which are measured in amplitude and rate, and the kinds are distinguished as numerically different rates of vibration. Thus classification is directly resolved into enumeration, and again number is kind. This is illustrated in the classification of light as colors which depend upon rates of vibration.

Something more than motion is manifested by the ether.

Light is the expression of ether as number and kind in the colors. Magnetism is the expression of space and form in position and direction. Heat is the expression of motion as force. Gravity is the expression of time as causation. Electricity is the expression of affinity as electrolysis. When the electric discharge is manifested by the electric sparks or the flash of lightning, it is manifested as light. Thus ether manifests the pentalogic concomitants both in quantitative and classific properties.

It manifests these properties by producing effects on ponderable matter, which effects appear to the senses and to the reasoning faculties as exhibiting quantitative and categoric properties; for example, light exhibits number to the mind, and when analyzed by the prism it exhibits color or kinds of light. Magnetism exhibits space relations in polarity and form relations in attraction. Heat exhibits motion in the particles of bodies as vibrations which may be increased in amplitude until the incorporation of the body is destroyed, when only space relations appear. Gravity manifests itself in pressure as continuous action, which appears as acceleration of speed in the falling body and as the cause of the fall. Electrolysis exhibits decorporation or the dissolution of the bonds of affinity, and reincorporation or the establishment of new bonds of affinity.

In the ethereal realm particles in inconceivable numbers coöperate in the production of effects in multiplied ways.

In the stellar realm the ethereal forces are found, for the stars exhibit the phenomena of light, magnetism, heat, gravity and electricity through the medium of the ether, not only because the ether surrounds the stellar bodies, but also that it seems to permeate them.

There are molecular forces believed to exist in stars as chemism; but the theater upon which their action may be studied is on the surface of the earth.

The forces exhibited in stars and in the systems of which the stars are particles are centripetal and centrifugal, as rotatory and revolutional. Gravity is a force which acts upon stellar bodies through a medium and which is transmuted into rotation and revolution and is again manifested in the figures of the bodies of the solar system, for they have the spheroidal form.

Thus the ether coöperates with the stellar orbs by transmitting light, magnetism, heat, gravity and electricity from one to another. These transmissions are made not by extracting them from one orb and transporting them to another as if they were bodies, but by inducing the motions in ether by which they are expressed, which in turn are induced by the ether in the body receiving them as an effect the cause of which is in the emitting body. In the language of the sciences of the ether the five ethereal concomitants are called radiant forces, but perhaps it would conduce to sound reasoning if they were designated radiant causations.

So also in the celestial realm body coöperates with body. The orbs of the solar system coöperate with one another in producing the solar system itself as a body, and they coöperate with one another through the medium of the ether in radiant causation as reciprocal cause and effect.

In the terrestrial realm the spherical bodies coöperate with one another in producing strains and stresses which induce chemical reaction, and thus are the cause of the special mode of motion which we call chemism. So sphere coöperates with sphere, formation with formation, rock with rock, molecule with molecule in the reincorporation of mineral substances, which is a reincorporation of forces as well as of forms.

Molecules of air coöperate with molecules of air in a wind. Molecules of water coöperate with molecules of water in a rain; molecules of air and water coöperate in a storm, while molecules of air, water and particles of dust coöperate with one another that vapor may be transformed into water antecedent to the storm. Molecules of water coöperate with molecules of water to constitute the stream, the current, the wave and the glacier, while molecules of rock coöperate in the boulder as it grinds its way, coöperating with other rocks in corrading the channel.

The terrestrial spheres coöperate with one another in all geological processes. Upheaval and subsidence with flexure and faulting are produced by the coöperation of the nucleus in yielding to pressure derived from the building of formations with material transported by the river, which was disintegrated by the action of rain which fell as storm blown by the wind caused by unequal temperatures induced by the ether caused by the heat of the sun. Endless illustrations can be given of coöperation in the terrestrial realm.

In the vegetal realm by the coöperation of protoplasmic particles chemical force is transformed into vital force and processes coöperate with one another in the same body. The process of absorption by the rootlets coöperates with the process of transportation to the leaves, and here they both coöperate with the process of transpiration, and these coöperate with the processes of osmosis in the redistribution of the materials to the growing parts, and these again coöperate with the process of assimilation where the growth takes place, and all of these processes coöperate with the process of reproduction by which the seed is formed.

Beside the coöperation in production above noticed, an additional coöperation is discovered in the higher forms, where individuals coöperate as sexes in their reproductive function. The vegetal forces coöperate with terrestrial forces in the disintegration of rocks into soils, in which function they also coöperate with chemism, gravity, and ethereal force. Thus coöperation in the vegetal realm extends throughout the universe.

In the zoönomic realm all other forces of nature

coöperate with the forces of animal life to accomplish motility. That the organism itself is a system of coöperating powers in which the function of every organ is necessary to the continuance of the function of the others is commonplace doctrine.

First we note that metabolism consists of two correlative systems, one of anabolism, the other of catabolism; that is, the one builds up, the other tears down. They are not only correlative, but to a large extent they are contemporaneous. In fact, there can be no building up without tearing down; that is, no placement which is not displacement, except that material may be stored adjacent to organs as fatty substances, to be used as needed after it has thus been stored.

In the animal body functional coöperation becomes still more efficient by more thorough specialization, when multiple like organs of like functions are eradicated.

In animal life the body is moved by the differentiated movements of its component parts. The body as a particle moves by impact from external influences in the higher incorporation of the earth, but it also moves as an individual by the differentiation of its own internal movements ideally determined. This force is motility as it is exhibited in all locomotion, by which we mean all motions of the parts of the body which are directly related to the environment by which the whole body or any part of the body may be carried from one place to another.

Through motility the property of judgment becomes the guide of the animal body, determining the movements of the parts, stimulating the function of one organ, inhibiting the function of another and causing them all to coöperate to a mentally determined end. Judgment thus controls function and through it produces locomotion, by which the parts of the body are changed in relation to one another, and by this power of changing the place of parts the power of changing the place of the whole is accomplished in the more restricted sense expressed by the term locomotion. Thus we make a distinction between vitality as a method of molecular motion and motility as a method of organic motion directed by opposing or correlative muscles, motility being thus directed by contraction and relaxation, which results in all forms of locomotion. In this manner food is masticated, swallowed, and moved along the intestinal canal and delivered to the circulatory system by appropriate muscles; then it is taken up by the circulatory system and moved by the heart with certain accessory muscles, and as the circulation proceeds excretory materials are discharged, all by appropriate muscles. There are also muscles for the movement of the limbs, all adapted to locomotion. Then there are muscles necessary for the reproductive functions and finally there are muscles for the movement of organs of sense. All of the motions thus indicated in a summary manner are the result of the forces which we call motility to distinguish it from vitality, which is molecular force. Motility is controlled by metabolism, and is the metabolism of opposing muscles where one contracts and the other relaxes. The reason for explaining contraction and relaxation in this manner was set forth in a previous chapter. We thus see how the processes of metabolism and motility coöperate. We also see that all of the organs of one system, as that of digestion, or that of circulation and excretion, coöperate. We also see that systems coöperate with systems. Finally it must be noted that all of the other systems coöperate under the direction of the nervous system, and are thus obedient to mind, being under the control of volition, which is choice of activity, which is the choice of affinity—the mutual selection of particles of matter for molecular association. If all of this reasoning is valid, affinity is molecular choice in the animate body, and we may hence conclude that all affinity is molecular choice, as it seems to be, for chemists who do not ignore affinity never find any other way of rendering the facts into language. So that affinity is practically synonymous with selection. I will to cross the street, I will to walk, I will to set the organs of walking in motion, and I accomplish it by controlling the affinities of molecules in metabolism. Thus a system of organs has been developed by which muscular metabolism may be accomplished by a constant supply of new material for anabolism, and a constant discharge of waste material by catabolism.

We cannot conceive or express these facts in any way except by teleologic concepts. It is now a fundamental doctrine of evolution that the organism is developed through the accumulation of effects by individuals in successive generations. Not that each individual in the hereditary line has such a concept of the future that he could foretell the ultimate result, but that he had such a concept of the immediate future that he purposely planned and executed immediate action, and while a perfect state was not known so that every action was the right action for the ultimate benefit of the race, yet the judgments of action were usually judgments of immediate benefit to be derived, and these judgments resulted in action, whether good or evil, for of necessity they were followed by action without waiting for their verification by experience. We have already seen that judgments of intellection do not become judgments of cognition until they are verified. Judgments of action result in immediate action, and are verified after the act only when they appear as sentiments of good or evil to control the will.

Let us see how this control of the functions is accomplished, and what part the different portions of the system take in the mental activities as they control the mechanical action. The brain seems to be the organ of mentation, but there are ganglia in the different mechanical organs which take a subordinate part in the general system of mentation in locally controlling motility by exciting or inhibiting activity. It is not necessary that the brain should deal with every muscle, but only with general ideas of action, while the ganglia control the details of activity, for there is a hierarchy of authorities which ramify to every cell particle in the system. Thus the brain has the means of inciting metabolism in every particle of the body. This is the machinery of habit by which customary actions are rendered apparently automatic. By an analogous process of reasoning we must conclude that every particle of matter in the system has judgment as consciousness and inference, and that these judgments are transmitted by the sensory nerves to ganglia in a collecting hierarchy which finally reaches the brain. The organs of sense sending their judgments by the sensory nerves from the exterior, the organ of feeling from the interior, and we are compelled to infer that every particle of matter in the animate body has judgment; and that in the organs of sense they have judgments of cognition, but in the mechanical organs they have judgments of good and evil. Then we may consistently infer that the ganglia are organs of conception, and we come back to the statement that the brain is the organ of mentation, which does not deal with judgments individually, but only with concepts.

In reaction animals coöperate with plants, rocks, orbs and ethereal bodies. The systems of coöperation of which we have made mention are developed into a higher sphere, and a new mode is discovered in human activities, as every man coöperates with his environment. We have seen how in motility the internal motions of the body are converted into external motions by deforming the body itself. By motility as expressed in locomotion, the animal body can change the relation of its parts, and thus of the whole body in relation to external parts, while all such changes of relation are in obedience to mind. The animal body, therefore, can move itself in relation to external bodies in a limited manner, and can thus impinge upon them and coöperate with external bodies at will.

In the coöperation of animal with animal, societies are organized. These societies are highly developed and best illustrated in human life. Men, through activities, coöperate with other men. We thus have a vast assemblage of coöperations, but in these activities the man must necessarily coöperate with plants, rocks, orbs, and ether as well as with other men.

The activities in which men engage are all designed to accomplish purposes, and in order that these ends may be reached, man invents by minute increments sundry agencies by which they may more adequately be reached. In order that we may understand this subject, consider the purposes to be accomplished. By a careful examination of all human activities and the purposes directly subserved, it will be found that they are naturally grouped in five classes. First, man pursues pleasure, and those things which give him pleasure are sought. These are the ambrosial, decorative, athletic, divinatory, and fine-art pleasures. Second, man naturally pursues welfare in length of life and abundance of health, and seeks to avoid disease and death. By these are produced those activities which are called industries. Under industries we have to consider kind, form, force, history and purpose. Third, for pleasure and welfare man has found it well to associate, and to promote these associations he finds it necessary to regulate conduct. He therefore naturally pursues justice, and that gives rise to institutions which are constitutive, legislative, executive, operative, and judicative. This leads to a fourth form of activity, which again divides into two forms, activity of expression and activity of reception. Thus it is that man invents languages which are emotional, gestural, oral, written, and technical. Fifth; but man in the pursuit of pleasure, welfare, justice and expression discovers that he makes many mistakes, and that pain, misery, injustice and misunderstanding are secured instead of the desired ends. He thus finds it advisable to pursue wisdom, and organizes the necessary agencies. Therefore these are the agencies for the increase and diffusion of knowledge, observation, acculturation, education, publication and research; for this diffusion it becomes necessary to teach and to learn; so research and instruction appear and become pursuits of life for wisdom. These pursuits of wisdom imply both teaching and learning.

Now, we have pleasure, welfare, justice, expression, and wisdom as the purpose of the five grand classes of activities. These activities are indissolubly associated, for it is found that one end cannot be accomplished without accomplishing them all. The act which is designed for pleasure becomes pain if it does not conduce to welfare; the act designed for welfare may decrease life if justice is not secured. The pursuit of justice may result in injustice if truth is neglected. The end pursued for truth may lead to error if wisdom is not reached. All of the permutations between pleasure, welfare, justice, expression and wisdom may be examined, and forever it will be found that they are indissolubly connected, and must coöperate in order that the end may be reached by the individual.

At the same time the individuals have their activities differentiated, so that the labors of every man are to a greater or less extent distinguished from the labors of every other man in every organized society. All of this differentiation of labor upon which the highest civilization depends illustrates in the most forcible manner the nature of coöperation, for society itself is organized upon the theory of coöperating activities.

In these activities men not only coöperate with one another, but they individually and collectively coöperate with nature, and thus external nature is made to assist man.

The club is but an instrument to coöperate with the hand to increase its efficiency. The flail is but a club with a handle for increased efficiency. The thresher is but a group of clubs placed upon a cylinder and made to revolve to increase efficiency. The snow-shoe is but an addition to the foot to increase its efficiency. The sled is but an improved snow-shoe. The wagon is but an improved sled, and the railroad train but an improved wagon. The lens is but an improvement to the eye, and the telescope is but an improved lens, and the microscope still another improved lens. There would be no end to the illustrations which could be cited to show the manner in which the arts of man coöperate with one another and with man himself.

CHAPTER XIII

EVOLUTION

We are now to consider what happens to particles with the passage of time. At the outset we must consider what it is that has persistence and change. The particle has five manifestations as five essential concomitants or constituents: unity, extension, speed, persistence, and consciousness. As all the concomitants inhere in one particle and the particle is unity, extension, speed, persistence, and consciousness, the concept of a particle not having all of these essentials is a pseudo-idea. If any one of them is taken away from a particle it is annihilated, for there is nothing else in the particle but these essentials. They constitute the particle.

By abstraction we consider essentials severally, and when we consider the relation of particles we still consider the relation of essentials severally. The relations of essentials are properties. One concomitant in one particle cannot be related to the same concomitant in another particle without a relation existing between the other concomitants of the particles; that is, there cannot be a relation of one unity to another without a relation of one extension to another, one speed to another, one persistence to another, and one consciousness to another, if the particles be animate. If we go on to consider persistence abstractly, we must still remember that the persistence is the persistency of a unity, an extension, a speed, and a consciousness. But between these persistent concomitants there are relations, and these relations are changeable; so when we consider persistence and change, it is the persistence of particles of essentials and change of relations of particles of essentials. Only relations are changeable, essentials are persistent.

By abstraction we consider essentials severally, and when we consider the relation of particles we still consider the relation of essentials severally. If we go on to consider persistence abstractly, we must still remember that the persistence is the persistence of a unity, an extension, a speed, and a consciousness.

He who cannot distinguish between concomitancy and relativity cannot follow this argument and cannot understand its fundamental doctrines. He who cannot follow up this distinction in all of its logical results under the conditions of complexity which are exhibited in the various bodies of the universe considered by scientific men, had better devote his time to metaphysical speculation where logical distinctions are confused and fine-spun theories of the unknown are the substance of philosophy; for scientific men deal with simple facts, though they may be tangled in relations, while metaphysicians confessedly deal in speculation about the unknown and boldly affirm that realities are fallacies. When a scientific man speaks of phenomena, he speaks of the manifestations of reality; when a metaphysician speaks of phenomena he speaks of manifestations of the unknown reality of which he dreams, while he deems that the realities of the scientific man are creations of fancy. In science all knowledge is verity and all fallacies are false inferences. In metaphysics all knowledge is illusion which manifests in a vague way an unknown reality.

If particles could exist without speed there would be no change and no motion; or if there was but one particle with speed there would be only rectilineal motion, but as there are many particles with speed they collide and deflect one another; deflected speed is directed motion. The first phase of directed motion is thus change of direction in free particles. Here we have persistence in speed and change in direction by which persistence is divided into portions by events of collision, and this manifests time. There could be no time without motion, and no motion without space, and no space without number. The first or simplest manifestation of time is the division of motion into parts by events. This gives us the simplest concept of time known to science, and whenever in science time is considered, some motion is divided into parts by events. Thus the motion of the earth about the sun is divided into annual parts by events, and the motion of the earth on its axis is divided into daily parts by events.

By the incorporation of particles into bodies the speed of the particle is divided into parts, one part of the speed inhering in the particle as internal motion, another part inhering in the particle as external motion of the body. The speed of the particle is composed of internal speed and external or corporeal speed.

In bodies we consider the corporeal speed, and one body may have greater speed than another, although one ultimate particle cannot have greater speed than another. It is the speed of one body measured in terms of the speed of another by which time is usually determined.

Particle speed is persistent or eternal. Corporeal speed can continue only while the body remains incorporate. So essentials are co-eternal in the particle, but are co-etaneous in the body.

When we consider the collision of one body with another we may consider the action as a force, and if the particles remain without change of incorporation, action and reaction are exhibited as mutual deflection. When we neglect the nature of this deflection we are considering the forces involved, but if we consider results and compare the paths of the bodies before collision with the paths after collision, we pass from the consideration of force to causation, for the cause of their collision was their incident paths and the effect of collision their reflected paths.

Thus the study of time as exhibited by bodies leads to the study of causation. So in causation we have more highly related time. If we consider relations of persistence and change in the particle, we consider it as time, but if we consider it in the body we consider it as causation. Time and causation are thus reciprocal.

There are different kinds of natural bodies besides the one ethereal body of all ethereal particles; (1) the celestial bodies of molecular particles; (2) the terrestrial bodies or spheres of petrologic particles, in which certain of the molecular particles are forever undergoing reincorporation; (3) vegetal bodies which are still more ephemeral and reincorporated from the mineral kingdom to exist only for a time and then to be returned to the mineral kingdom; (4) animal bodies which are incorporated from the vegetal kingdom; (5) societies which are ideally incorporated.

In this incorporation they exhibit successions of causations; but causations are processes, and one abstract process cannot exist without the concomitant processes—that is, there can be no processes of causation without processes of force, form, and kind, together with processes of mind.

We know little of the reincorporation of stars, but we know much about the reincorporation of rocks, plants, animals, and societies. The study of incorporation and reincorporation is evolution from the standpoint of causation, which in turn is the study of time.

The consideration of the totality of changes occurring in the universe is evolution. These changes can all be resolved into changes in the position of the ultimate particle of matter. Directed changes in position lead to incorporation, then incorporation is succeeded by reincorporation, and the totality of these changes is the totality of evolution. Starting with this concept we may be able to redefine evolution in a more satisfactory manner at a later stage.

If there were no motion there would be no time but only persistence. If there were no incorporation and reincorporation there would be no evolution but only time as it is exhibited in the ethereal particle. At the very outset, then, we have to consider incorporation in the association of one chemical particle with another.

The theater of the motion of every ethereal particle must be circumscribed by the theater of the adjacent particles; we are logically prohibited from any other conclusion. When particles unite with one another in constituting a body, so that the speeds are divided into internal and external motions, by virtue of the external motion, they may change their space relations to external particles. In order that there may be bodies with changeable environment the particles must, by some means or another, associate. The first cause or method of evolution is choice; the first effect of evolution is change of environment. From this datum point we may go on to discuss the evolution of the laws or methods of evolution.

Affinity is choice of association in atoms and molecules by which new kinds are developed by the development of new orders of units through their incorporation into one body. This is illustrated in the conventional numbers where the ten units of one order constitute one of a higher order. Again, the bricks of a house, thousands in number, constitute one house in a body of a higher order. Now the atoms of a molecule are associated by affinity, which surely resembles choice of association, though it may be the choice of dominant particles or mutual choice; but in the bricks which constitute the house their association is the choice of volition in the builder, by the choice of activities in the control of his muscles. This choice of activity still relates back to a choice in the reciprocal processes of metabolism, which again is affinity. Thus external choice is controlled by mind through internal choice.

The primal law of evolution seems to be psychic. We shall call it the law of affinity and define it as the choice of particles to associate in bodies. The facts observed in the chemical incorporation of particles into bodies are explained by this hypothesis, but they remain the same whether the explanation be valid or invalid, that is, whether we consider affinity to be due to psychic choice or to some unknown mechanical property.

By the incorporation of atoms into molecules particles become bodies which react in collisions with the environment in a new manner; thus bodies can perform functions which particles cannot perform. This leads us to the consideration of incorporation as organization, when functions and organs as concomitants are transmuted together. Molecules, because they are incorporated numbers, are organized numbers, or in other terms, chemical organization by incorporation is numerical organization. Now we must see what these new functions or reactions are.

Molecules of substances are aggregated into stellar bodies by their mutual reactions through the gravitating medium—the ether. Thus a second method of evolution is developed which is known as adaptation to environment. By this method not only are the celestial bodies incorporated into higher units, but their forms are subsequently controlled by the same law when they yield to the forces of the environment as spheroidal figures, rotating and revolving as fluid bodies. Stars are evolved under the law of adaptation to environment and remain under its control in their changing figures through the history of their revolutions.

Under this law stars change their environment, passing through a succession of positions in a cycle of revolution. This seems to be a valid statement of the changes brought about by the incorporation of atoms into molecules, and their further incorporation into stars, and their still further incorporation as stars into systems.

In celestial bodies we know only of the fluid state of matter as revealed by astronomy. While there may be solid bodies in the other orbs, as in the earth, astronomical investigation does not reveal them to research as solids; such solid bodies are recognized to be studied only in the earth, where they are revealed as rocks, and if they may exist in the other orbs the science of astronomy does not deal with them. The forms of the stellar bodies are those assumed by fluids under the force of gravity. The stars themselves are particles in systems which are bodies of a higher order. Events are discovered in the motions of the celestial orbs and exhibited in a great variety of ways as set forth in the science of astronomy.

Thus states of motion are divided into events of motion. The states are represented by rotation of body, which is the revolution of particles, while events are marked by phenomena which attend the rotation and revolution. These are phenomena of time, or persistence and change. Then the heavenly bodies are constantly changing their relations to one another, and a vast system of perturbations are discovered. Motion at apogee differs from motion at perigee, motion at aphelion differs from motion at perihelion, and a great variety of perturbations of path are revealed. Here we study causation. Finally, the genesis of the heavenly bodies is studied as their evolution.

LaPlace was the founder of this department of astronomy. The researches in this realm had revealed the common direction of motion in the orbs of the solar system, the small eccentricities of path, the inclination of the orbits, and the conservation of areas. Reasoning that contraction would accelerate rotation and hence oblateness, he conceived the hypothesis of the evolution of the solar system on the theory of the radiation of heat into solar space from a nebulous mass. He conceived that this mass, revolving in an orbit, constantly accelerating and thus constantly increasing its oblateness, would thus gradually retire by attraction from an external ring of matter which would ultimately break up into one or more orbicular bodies.

Since the time of LaPlace his method of accounting for satellites as a breaking up of rings has been questioned, and facts have been discovered that give ground to the conjecture that other methods of separation into bodies by fission are not only possible but even probable. This new doctrine arises from the investigation of binary stars. It will be observed that LaPlace's theory was an attempt to harmonize many diverse laws discovered by induction and verified by deduction, by accounting for them all by one fundamental doctrine of evolution, which is no other than the adaptation of every particle of matter to the conditions imposed upon it by every other particle in the environment. Under this hypothesis LaPlace promulgated a doctrine of evolution which, in its fundamental elements, has remained to the present, notwithstanding the tests of observation and recomputation to which it has been submitted, though minor components of the doctrine are questioned.

There is still another assumption of LaPlace that must now be questioned, as it is unnecessary to his argument and incongruous with facts herein demonstrated; his assumption is that heat is radiated into space and that it leaves the cooling body to join external bodies. All of this was quite compatible with the concept in vogue in his time, when heat corpuscles were supposed to be itinerant from body to body. Now we know that heat is not a special form of matter, but is only a deflection of the motions of the particles of matter whose speeds are constant, and that one body causes heat in another but does not yield heat as speed of particles so that it loses what the other gains. While the heat of one body induces heat in another, no motion as speed leaves the cooling body, but its reaction transmutes the heat motion into the structural motion of the body and that reaction which we call the transfer of heat from one body to another is in fact its equilibration through mutual transmutation.

Thus, by the theory of LaPlace, the chemical changes proceeding in the combination of atomic particles existing in the nebulous mass were accelerated by gravity until they were consolidated into stellar bodies, the process being a succession of recombinations in molecules of higher orders.

In the geonomic realm three so-called states of substance are found: the ethereal, the fluid, and the solid. All of these states are conditions of incorporation. Gases may become liquids and liquids may become solids, and vice versa, by progressive incorporation and reincorporation. These states of substance often exhibit interesting critical points in which secular changes are accelerated by sudden metagenesis, especially at critical points of temperature and pressure. Thus changes of state are secular metageneses accelerated in sudden metageneses. Everywhere and forever the states are changing by events and the geonomic realm is forever in flux. The winds are in motion, the waters are in waves, tides and currents, and the waters themselves are evaporated and move in clouds through the air and are condensed into streams that flow into the great bodies of water and into the ocean itself. The fluid waters are transformed into solid, and the solid are gathered at high altitudes and high latitudes into great bodies of ice that are forever growing, melting, and moving forward. The solid rocks are forever undergoing geologic changes under the stress and strain produced; thus molar metamorphosis is forever in progress. The rocks are carried from the land to the sea and the sea-bottoms are upheaved, while mechanical changes are forever in progress throughout the solid envelope. States appear to be changed into other states only by events which come in winds, storms, earthquakes, and fires.

That which we are to note as germane to this argument is that there are three states of matter involved in the study of geonomy: the ethereal state in which the phenomena of heat and electricity are observed, the fluid state, and the solid state in which the especial phenomenon of the geonomic orb—the earth —is observed. As in the stars we are compelled to discuss ethereality, terrestrial heat, light, electricity, magnetism, and gravity, together with centripetal and centrifugal force and fluidity, so in the geonomic realm we must study not only the same subjects, but must also consider the solid condition with the stresses and strains involved and the metageneses that appear through chemism.

In the ethereal realm we know of the ethereal state; in the stellar realm we know of the ethereal and the fluid states; in the geonomic realm we know of the ethereal, the fluid, and the solid states.

In the study of the earth a differentiation is found in the air, the sea, the land, and the nucleus. They are also integrated by the rotation of the earth, which is the revolution of its particles. The air is imperfectly differentiated into winds. The waters are differentiated into seas with gulfs, lakes with bays, and rivers with creeks, brooks, and rills. Then the waters are evaporated and differentiated into vapor, and these vapors become clouds and the clouds become rains. Then the waters that were evaporated into vapor and condensed into rain are also frozen into snow and ice, and ice itself plays an important part in the mechanical changes wrought upon the surface of the earth. Then the solid sphere is differentiated into formations, and the formations into rocks or blocks, and these again into crystals and grains; then the rocks are ground by the running waters and blown by the winds and distributed through the air and over the land as dust. They are also carried by the waters into the sea and deposited in formations, and finally they are carried in solution by the interpenetrating waters into the crevices of the rocks, by which blocks are parted. Finally, fluid masses from the molten interior are thrust into the rocks in dykes, chimneys, and lacolites, and spread over the surface in coulees, cinders, and dust. All of this commingling of materials results in a recombination of substances ever found to be more and more highly compound. At the surface of the earth these changes are still further multiplied in the production of soils, which is accomplished by the wash of rains, the grinding of ice, the chemical decomposition of the rock, especially aided by heating and cooling, together with the disintegration that arises from the action of plants and animals upon the soil, and by the commingling of their bodies with it, so that a highly compound mass of particles is produced, known as the soil. This soil is the theater of chemical changes by which the more highly compound molecules are developed, necessary directly to vegetation and indirectly to animal life. As chemical compounds are more sensitive to change, mineral forms are more sensitive to metamorphosis, and as mineral and molar forms are changed processes are multiplied and become more efficient in the production of change. Thus the new law of evolution which we find in the geonomic realm, is the acceleration of change by increasing heterogeneity. It may be called the method of heterogeneity.

The law of affinity and the law of adaptation found in the astronomic realm also pertain to the geonomic realm. But to them there is added this new law of heterogeneity. Thus an incessant metalogosis, metamorphosis, and metaphysisis results in universal, constant, and multifarious metageneses.

As substances become more compound they become less stable, and acceleration of heterogeneity is the acceleration of metagenesis.

In the phytonomic realm, that is, in plants, a fourth state of substance is found. This is the vital state, for plants have life. Substance in the fluid and solid states is taken up by the plant through the medium of the ethereal state exhibited in light and heat and metagenetically changed into the fourth state as vitality. These metagenetic changes are known as assimilation, by which the plant is produced. Plant growth is secular, and the materials pass through the fluid state into the living state, which is growth; the plant may then dissolve secularly by decay, or by sudden change in combustion.

We cannot understand the plant without a consideration of all the four states of matter and all the four changes of matter which occur therein as events. As the plant grows, minute molecules are added; as the plant decays, minute molecules are taken away, as the vital changes observed.

Vitality as a state first finds expression in the continued growth of the plant, and a still higher expression in the heredity of the species, for the state is continued from plant to germ through the germ in life and growth to reproduction, where it again appears in the new germ. Thus we are compelled to consider the vital conditions of heredity. The metagenetic changes of the individual are bequeathed to its posterity, and the environmental changes of the individual are wrought into its structure and these again are bequeathed within more or less restricted limits. Thus in the study of the plant we study a new state of substance, and new changes are here events in the history of the individual, transferred by heredity to its offspring. In the consideration of the development of germs into adult individuals, the accomplishment of the process is ontogeny. In the consideration of the development of individuals in generations by which the race is evolved, we may consider the result reached as phylogeny.

In this realm the law of the acceleration of evolution is the one discovered by Darwin and known as the survival of the fittest in the struggle for existence. Plants multiply by germs, and more germs are produced than can possibly find room on the surface of the earth when developed into adults. Plants multiply by hundreds, thousands, hundreds of thousands and perhaps even millions; some must perish by inexorable conditions, and the few that arrive at maturity are those best adapted to the local environment where they live; but the germs themselves have their environments changed by mechanical agencies, as on winds, waves and streams, and they are often carried about by animals. This change of environment modifies the plants themselves in such a manner that varieties are developed which ultimately become species.

The evolution of plants is fundamentally chemical under the law of affinity; it is accelerated by adaptation to environment; it is then subject to the law of acceleration by heterogeneity, and evolution is still further accelerated by the survival of the fittest.

In animal life a fifth state of substance is found which I call motility. In this state all the other states are found: ethereal, fluid, solid, and vital. Changes which occur as events in the history of motility are collisions and metageneses in the ethereal, fluid, solid, and vital states, but to them is added a series of changes which are expressed in motility. In the animal, anabolism and catabolism are contemporaneous, as the animal has coeval growth and decay; anabolism and catabolism then become metabolism. The dual processes of metageneses, which are evolution and dissolution, are now combined so long as life lasts. In rest, and especially in sleep, anabolism may progress at a greater rate than catabolism; in exercise, and especially in violent exercise, catabolism may prevail, but neither can wholly cease while the motile state endures. Here, in the state of motility, ontogeny appears in the individual and phylogeny in the race.

In attempting to define the states of substance a precaution is necessary. It must be understood that the states of matter do not always appear to be separated by hard and fast planes of demarcation; and so far as we can assert with confidence, there seems to be a gap between ether and ponderable matter, though a complete recognition of the ether is but an event of the present day. The gaseous and liquid states are included in the fluid state. Between the fluid and the solid states intervening conditions are found.

It is known that no perfect distinction can be made between the solid and the vital state. There are those who believe that an impassable barrier exists between them, but this doctrine is rapidly being dispelled; indeed it is no exaggeration to say that scientific men rather confidently believe that the barrier is soon to be thrown down. That the barrier between vitality and motility has been overthrown is believed by many biologists, though there are still those who believe that the apparent consciousness of plants as exhibited in a great variety of phenomena can be explained as mechanical phenomena. If this lingering belief be true, the barrier still exists; but there is no ontogenic barrier even if there be a phylogenic barrier.

In the motile state of matter the special law of evolution was discovered by Lamarck. It is the law of effort, and may be stated as the development of organs by exercise and their extirpation by disuse. It must be remembered that all the other laws of evolution apply to the animal and that this new law is added in the motile state. Sometimes the law of heredity is called a law of evolution, but in fact it is the law of the continuation of species both vegetal and animal, and is not a law of evolution.

In the animal the law of affinity still appears in metabolism as fundamental, for by metabolism the development of the organ is accomplished and without it there could be no effort.

The animal is adapted to environment by many ways, especially in the development of agencies for accommodation to climate, as in the down of birds, the fur of animals, and in various protective devices as external coverings exhibited in the shells and shards of the lower animals. But the animal adapts itself to environment in another manner: endowed with locomotion, it seeks a favorable environment best adapted to protection and best adapted to supply stores of food.

The animal is still subject to the law of heterogeneity, for the multiplication of heterogeneous characteristics adapts it to heterogeneous conditions of environment, and so the limitations to the multiplication of adults are largely thrown down.

The animal also is subject to the law of survival, for notwithstanding the utilization of every possible environment for every variety, there is yet an overmultiplication of individuals, which must perish.

Upon these laws supervenes the law of effort by which organs are developed on various lines for various conditions of environment, and the result of this organic evolution leads to the survival of those best adapted. In animal life evolution is by affinity, adaptation, heterogeneity, survival, and effort. The first of these methods is the basis while the others are successive accelerations, so that the changes wrought in the animal are progressive in geometrical ratio by the compounding of all the factors.

Another factor in evolution appears in the organization of demotic life which may be observed among those of the lower animals in which societies are found, appearing among mankind and becoming the chief factor in civilized life. It is a method of evolution to which inadequate attention has been given, and the failure to recognize it has led to misapprehension of the nature of human evolution and to preposterous claims for the efficiency in mankind of the laws of evolution found among lower animals. This mode of evolution, therefore, needs more elaborate presentation than that which we have already given for the other laws. By man in civilization the law of effort is transmuted into the law of culture, the method of invention; that is, the effort is designed effort for the purpose of improving human conditions. The chemical law still remains valid, but the exercise of organs is ever from age to age, century to century, and even decade to decade concentrated upon one special system of organs. Of the five systems, that which has the function of thought and which is the nervous system is ever more and more exercised, until metabolism itself is accelerated to such a degree that the changes in the nervous system are far more rapid than in either of the other four systems. Thus human evolution comes to be mental evolution, and this mental evolution is the product of culture by invention.

The law of culture transforms and then absorbs the law of adaptation, the law of heterogeneity, the law of survival, and finally the law of effort. In what manner this transformation and absorption are effected must be explained. In man adaptation to environment is transmuted into the adaptation of environment to man. Man is not adapted to food, but food is adapted to man by culture. New foods are developed until many are used. The animals which furnish food are cultivated and multiplied under the direction of man. Vegetal foods are in like manner multiplied and cultivated in vast fields, vineyards, orchards, and gardens, and new varieties are forever developed by the skill of man.

Man is not adapted to the environment of climate, but he adapts the climate to himself; when it is too cold he kindles a fire, and he protects himself when away from the fire by clothing; when it is too wet he covers himself with a roof; when it is too windy he protects himself with walls; thus man does not develop down like the birds, or wool like the mammals, or carapaces like the turtles. Man does not develop fins for life in the water, but he constructs boats that he may dwell on the sea. Man does not become a climber to live on the trees, but he ascends the trees on ladders and he fells the trees for temples. Man does not seek shelter among the rocks, but he quarries the rocks and builds palaces. Man does not burrow in the ground, but he molds and burns the clay and constructs marts of trade. Man does not develop eyes that he may live in the dark, but he invents lightning light that night may become day. The illustrations of the change of adaptation from man himself to the environment may be found in endless profusion. There is a change wrought in man by all these agencies, but it is a change in his mind exhibited in the development of the organ of mind and the concomitant development of thought.

The law of heterogeneity undergoes a like transformation. Upon the things in the environment which are useful to man the law of heterogeneity is concentrated. Domestic animals are multiplied in variety, and cultivated plants are changed until their native forms are lost and the new forms are multiplied beyond enumeration. Fabrics for clothing are produced and garments are made; materials for house structure are differentiated from the materials of nature, and dwellings, storehouses, marts, and temples are constructed in a multiplicity of forms. Tools and machines are differentiated from natural material; all the powers of nature are specialized for man's purposes; the whole progress of mankind is a succession of differentiations or specializations of the materials of nature to become the works of art.

The law of survival also undergoes a profound modification. Men are no longer subject to the vicissitudes of natural environment where winds may congeal their limbs, where floods may overwhelm them with death, and where disease may carry them away in multitudes. These agencies still act and have their victims, but the inventions of man are ever becoming more potent for the preservation of life. There was a time when drought in a narrow belt of country might produce a famine and when the people of such regions might perish; but no more famines can occur, for railroads link all fields to every man's farm. There was a time when a blizzard might destroy a tribe; but now a storm may sweep in vain from the boreal zone about the dwellings of civilized men, for man constructs his home against these vicissitudes.

Human providence is more potent than flood, more potent than drought, more potent than wind. The man of intellect wields a power that giants cannot exercise.

The differentiation of animal species found in the lower world is replaced as a new method of progress is evolved. The animals differentiate into biotic species. This tendency seems to have prevailed in the early and more animal history of mankind, for the records of these forms are still preserved in types of men, as exhibited in the conformation of the skeleton and especially in the cranium; it is also exhibited in the color of the skin, the structure of the hair, the attitude of the eyes, the conformation of the face, and in other ways. But there is no black, or white, or tawny species, there is no straight or woolly-haired species, there is no horizontal or oblique-eyed species, there is no blue-eyed or black-eyed species, there is no broad or long-skulled species, but these characteristics are now intermingled in inextricable confusion—the result of the admixture of streams of blood. Thus the method of differentiation of animal species has been reversed in the case of man. That in which men now differ is intellectual power, but fools are not necessarily blue-eyed and wise men black-eyed. The traits in which men differ are moral, but honest men are not necessarily broad-skulled or rogues long-skulled.

The law of adaptation in the lower animals and in plants was made efficient by a high rate of multiplication, but in civilization this rate is diminished, so that man has not even yet crowded the earth and no land has been inhabited so densely as to press upon the capacity of the land to produce food. Famines have occurred only by improvidence, and the poor starve by neglect. The effort of mankind for sanitation, the healing of wounds and the curing of diseases, is the endeavor of mankind to repeal the law of nature when the environment is his destruction; thus this law of adaptation to environment for the preservation of the few among the lower animals, is made inefficient by the slow rate of the multiplication of men and is replaced by human effort to preserve and prolong life.

There is an environment to which men are adapted; it is the environment of culture. Most men speak the language of the people among whom they were born. Every man seeks a vocation to adapt himself to the vocations of others, that by his special labor he may acquire the most of the special labors of others; so he adjusts himself to the industrial conditions by which he is surrounded. From the cradle to the grave his intellectual advancement is dependent largely upon his intellectual environment, and he suits that environment to his purpose.

Man cultivates his physical powers by exercise in the industries and in a variety of athletic sports, in the same manner as do the lower animals, and he invents new methods of physical training; but he also and chiefly develops methods of intellectual training, instruction and research, to which the schools, the libraries, the journals, and the systems of research abundantly attest. No, the laws of brute evolution have been repealed by substitution and the new ways are methods of culture. The laws of nature unmodified by man produce horns, claws, fangs, and poisons for attack, with armor, cowardice, and deceit for defense. Culture replaces these brutal devices; smiling fields, cheerful homes, and all the products of civilization are derived from the inventions of man himself. As the generations come each inherits from his predecessor and adds to the patrimony by self-activity. That which the self can accomplish is multiplied by all which the social environment produces. Man is not only an heir to the past generations, but he coöperates in the activities of the present, and when he dies he bequeaths the intellectual wealth which his self-activity has acquired in coöperation with all his contemporaries of the world.

In the natural world evolution is primarily by incorporation and reincorporation. This incorporation is by affinity. We have shown that affinity is explained as the consciousness and choice of ultimate particles. When we reach animate beings in which affinity is mind, the Lamarckian law of effort becomes potent in evolution until men are developed and the five forms of culture are invented. Molecular reincorporation by heredity now goes hand in hand with culture or self-activity modified by social environment.

Evolution as a process is the development of new kinds with their concomitant forms, forces, causations and ideations by the multiplication of the relations of causation.

CHAPTER XIV

SENSATION

We must here recall the distinction between feeling and sensation set forth in a former chapter. Feeling is cognition of effect upon self and gives rise to the emotions, while sensation is the cognition of the external cause of a sense impression and gives rise to intellection.

I feel light as an effect, but I see its cause in the luminant or the reflector. I attend to its effect, if it is too brilliant, or I attend to its cause, if I am interested in the cause. When I attend to the effect, it is a feeling; when I attend to the cause, it is a sense impression. An explosion occurs; the effect upon my ear is painful. If I attend to it I have a feeling, but if I wish to know its cause and attend to it, I have a sensation. Thus feeling and sensation are reciprocal. The more the feeling, the less the sensation; the more the sensation, the less the feeling. This is an old doctrine in a new form. The old doctrine of psychology is this: that feeling and cognition are inversely proportional; as we go on the old statement will be found faulty, and the new statement, that feeling and sensation are reciprocal, will be found correct.

An object impinges upon my organ of taste. If its taste is pleasant or unpleasant and I attend to that as an effect, it is a feeling, but if I attend to the object as a cause, it is a sensation. The organ of taste is in the portal to the metabolic organs. The taste of the object is pleasant or unpleasant, as I perceive by eating. Now, suppose that I am selecting apples for the purpose of putting the sweet in one basket and the sour in another. I am attending now to a property of the object through its effect upon the subject. Its effect upon the subject is emotional, but considered as a property of the object, it is intellectual. It is thus that an organ of feeling is transmuted into an organ of sense which reveals the property of the body.

The feeling of the circulation, which is variable by temperature and thus a feeling of heat, is developed into the sense of touch, and the sense of touch, which reveals the property, performs the vicarious function of revealing the body touched.

The feeling of strain is developed into the sense of stress, and the sense of stress reveals the body producing the stress.

A feeling of vibration occurs when the medium, as water or air, is agitated in such manner as to produce sound. This feeling is especially produced in the self by speech; the origin of speech is the calling of the mate, which call is made by one and heard by the other, and hence heard by both.

Thus the feeling of sound develops into the sense of hearing, which is the sense of causation; for the primordial ego, in the race and in the individual alike, first cognizes causation as speech and distinguishes it from force, for it can cause another to act by speech and it is conscious that it can be caused to act by speech.

The feeling of motion in self results whenever we are conscious of the will to move. Thus the will to move is the cause of the feeling of molar motion itself, and the feeling of motion is developed into the sense of vision by which motion is primordially and naively interpreted as the sense of conception. The feeling of motion is developed into the sense of seeing, for we feel molar motion and feel that that is caused by will, and primitive man naively infers that all molar motion is caused by will; hence he infers that all molar bodies have will.

The senses are vicarious feelings.*

I have already defined consciousness in the particle as awareness of self, as a unit, an extension, a speed, and a persistence, for this is the hypothesis upon which I am working. For human psychology it

---

*In this work only such a review of science is intended as is necessary for the development of an epistemology. In order to accomplish this I have attempted to set forth the properties of bodies in their reciprocal aspect as bodies and particles, or as internal and external relations. I have not considered it necessary or appropriate to enter into a minute discussion of the facts and principles of all the sciences severally. For example, the development of the senses from the feelings receives but brief mention. To set forth the ontogeny and philogeny of the senses would require a separate work. In my consideration of the development of the sense of hearing I have followed Frederic S. Lee, more perhaps than any other physiologist, though I have consulted several other authors on the subject. In stating my conclusions I have necessarily refrained from citing authorities, as I do not enter into these subjects except to make broad generalizations. But since this chapter was written I have received an abstract of a paper read by Dr. Lee before the British Association (published in the Report of that Association for 1897), in which I find that he briefly but clearly propounds the doctrine that the feeling of equilibrium is developed into the sense of hearing. I quote the abstract in full.

THE EAR AND THE LATERAL LINE IN FISHES.

BY FREDERIC S. LEE, PH.D.

The chief morphological facts upon which the theory of the origin of the ear from the system of the lateral line is based are similarity in structure of the adult organs, in innervation, and in ontogeny. Physiology seems able to present at least circumstantial evidence in favor of this theory. The author has investigated the functions of the ear and the sense-organs of the lateral line in fishes.

The Ear.—The results may be tabulated as follows:—

Functions of the Ear.	Sense-organs.
I. Dynamical functions in recognition of.......	1. Rotary movements. Cristæ acusticæ 2. Progressive movements. Maculæ acusticæ

needs not that the theory be extended to the ultimate chemical particle, but the doctrine is demonstrated to the extent that the animate cellular particle is conscious. Now I wish to consider consciousness as awareness of the part which a particle takes as a cause or effect in the production of a judgment.

When a sapid substance impinges upon my organ of taste I am conscious of an effect. When a body touches me I am conscious of an effect. When a sound impinges on my ear I am conscious of an effect. When a body presses upon my muscles I am conscious of an effect, and when a color strikes

---

II. Statical functions in recognition of...... } 3. Position in space. Maculæ acusticæ.

The above functions are divisions of the general function of equilibration: the sense-organs of the ear deal with the equilibrium of the body under all circumstances, both in movement and at rest.

In vertebrates above the fishes we must add to the above :

III. Auditory functions in recognition of. ..... } 4. Vibratory motions. Papilla acustica basilaris.

Experiments by the author and by Kreidl prove that fishes do not possess the power of audition. Hence the ear in fishes is purely equilibrative in function.

2. *The Lateral Line.*—Simple cutting of the lateral nerve or destruction of the lateral organs does not seem to affect equilibrium. But destruction of the organs, combined with removal of the pectoral and pelvic fins, causes marked lack of equilibrium, manifested by uncertain, ill-regulated movements; removal of fins alone has no pronounced effect.

Central stimulation of the lateral nerve causes the same compensating movements of the fins as does stimulation of the acoustic of the opposite side. These results make it probable that the organs of the lateral line are equilibrative in function, and are employed in the recognition of currents in the water and of movements of the body through the water. The results of Bonnier and of Fuchs are in harmony with this.

This was probably the primitive function. By the inclosure within the skull of a bit of the lateral line and the differentiation and refinement of its sense-organs, a more perfect organ of appreciation of movement, and hence of equilibrium, was evolved in the ear. Along with the appearance of land animals a portion of this organ became still more differentiated and refined, and, as the papilla acustica basilaris, acquired the power of appreciating the movements that we call sound. Thus equilibration and audition became associated in the same organ.

my eyes I am conscious of an effect. In these cases consciousness in a judgment is awareness of effect on itself, but it is the consciousness of the particle which is transmitted to the cortex.

See how this is developed. Consciousness is awareness of the part which self takes in the production of a judgment, either as a cause or as an effect. Thus I am conscious of the cause when I act upon another, and I am conscious of the effect when another acts upon me, and I am conscious of both cause and effect when I act upon myself, as when I touch my head with my hand. Here there are two pairs of correlates, self and other, together with cause and effect, and we must distinguish an active consciousness from a passive consciousness. I call it an active consciousness when I am conscious of being a cause, and a passive consciousness when I am conscious of experiencing an effect. This distinction must be firmly held.

Consciousness in this stage is awareness of the terms of causation, but they are not immediately related, for cases of active and passive consciousness occur usually at different times and under different circumstances. But there are some occurrences where the active and the passive elements are immediately connected in succession; this happens when I act on myself. In this manner the primitive mind learns of causation as composed of cause and effect, in the order of antecedent and consequent.

When I am conscious of an effect I infer a cause as an external object. When I taste I infer that I taste some other thing or object; when I smell I infer that I smell some external thing; when I am touched I infer that I am touched by some external thing; when I am pressed I infer that I am pressed by some external thing; when I hear I infer that I hear some external thing, and when I see I infer that I see some external thing. This something we call the object, and the mental act we call inference. A consciousness and an inference produce what I call a judgment, but this is an imperfect account of the process; let us know it all.

A sense impression does not constitute a sensation, but a sensation is compounded of sense impressions. Let us say that I have had many sense impressions of different kinds. Now suppose that I have one of taste; how shall I classify it with former sense impressions? Evidently they must be recalled and compared, and I choose one for this purpose. This choosing of a past sense impression and comparing it with a present sense impression and deciding that they are alike, I call a judgment.

These things are necessary to a primitive judgment. First, a sense impression; second, a consciousness of that impression; third, a desire to know its cause; fourth, a choice of a cause; fifth, a consciousness of the concept of that cause; sixth, a comparison of one conscious term with the other; and seventh, a judgment of likeness or of unlikeness. Stated in another manner, the judgment has these elements, a consciousness of a sense impression on the one hand, and a consciousness of another which is chosen, and the two are compared and found to be alike or unlike as the case may be, and a judgment is made. In still another manner a judgment may be defined as the comparison of a present event with a past event in which consciousness is twice involved; in the first an impression causes consciousness; in the second a choice causes consciousness, when the two are compared and a judgment made of likeness or unlikeness, which is identification and discrimination.

Choose a taste and you will recollect a taste; choose an odor and you will recollect an odor; choose a touch and you will recollect a touch; choose a pressure and you will recollect a pressure; choose a sound and you will recollect a sound; choose a color and you will recollect a color.

To choose is to revive in memory, for choice is the cause of the revival which is the effect. You cannot think of all sense impressions or sensations at one time, but choose any one of them and you will recollect that one. Inference, therefore, is guessing or choosing, or in another light it may be called interpreting. We shall hereafter see that this choosing is not random guessing.

The babe tastes milk; tastes it again and makes a judgment that the milk which it tastes now is like the milk which it tasted before; then it tastes vinegar and makes a judgment that it is unlike the previous taste. It continues to taste milk and vinegar and discriminates between the two. Its judgment of likeness is repeated in the case of the milk and repeated in the case of the vinegar and these judgments are consolidated, so that the present judgment of likeness is a judgment of likeness to some of the previous cases and of unlikeness to others. The mind does not recall every example to consciousness and compare them severally with the present one, but it recalls the like in a consolidated or fused group if the judgment is that of likeness.

This process of consolidating or fusing judgments I call conception.

It has been said that an inference is not a random guess. The guess is always dictated by something in experience as some collateral circumstance, expectation, or interest. We shall hereafter see that interest is the chief, if not the sole, agency in determining the choice.

But some judgments are not valid. A taste may be subjective, due to some disease of the organ of taste; then the judgment is a fallacy.

Suppose that my skin is diseased, and that I have a feeling which I mistake for a sensation and infer that something touches me; this subjective effect, which I here call a feeling, must be distinguished from a sense impression or it will lead to an erroneous judgment. I may have a feeling in the ear, as when I take an overdose of quinine, and if it is confounded with a sense impression a fallacy is produced. Feelings of this kind are sometimes known as subjective sensations, and they must always be clearly distinguished from sense impressions.

Here we reach a dilemma; a judgment has been formed, but it may be a fallacy or a certitude. How shall we know? Something else is needed; this is verification. In sensation verification is accomplished by repetition. But this is an imperfect method, for in abnormal conditions repeated erroneous judgments may be made. While the method usually serves the purpose, sometimes it fails and a higher verification is dependent upon another faculty of judgment by another sense.

Verification depends upon the ability of a judgment to coalesce with other judgments in concepts; that is, it depends on its conceivability. If a judgment is incongruous with previous judgments it cannot be conceived and is held for confirmation or rejection. The class may at once be discovered and the right concept enlarged, or it may wait until another like judgment is made, when a new concept will be generated.

Primary consciousness is in the end organ, but it is transmitted by fibrous nerves to the ganglion and finally to the cortex; when it comes to the cortex, the individual, or the ego, is conscious of the same impression. Each ganglion in the hierarchy forms a distinct judgment. The cortex certainly forms judgments for itself and combines them with concepts. The action of the cortex must be concomitant with the making of a judgment, and as the judgment must coalesce with the concept, the part of the cortex involved must be structurally modified thereby. Thus it is that a record is made of a judgment when it coalesces with a concept. The record then is physiological, as memory is physiological, and judgment and conception are thus the psychological abstracts of concomitant processes of the brain.

A judgment once formed remains in memory as an effect on the organ of mind; another like judgment revives it, or in more common language, it is recollected. Memory as retention is not a phenomenon of the fifth property called judgment, but of the fourth property called time; but recollection is revival in consciousness and is an intellectual process. To distinguish the fifth property from all the others we may call judgment intellectual and the other properties mechanical. It must be remembered that the judgment cannot exist without the mechanical properties, that is, there can be no judgment without retention or memory. A judgment cannot persist as a pure judgment, for its duration, which is called memory or retention, depends upon the time property of a body which must also have motion, space, and number.

In experimental psychology the mechanical concomitants are the units with which judgments are measured. The science also deals in experiment with the conditions in the object under which judgments are formed. It may be that here it finds its most fruitful field as a co-worker with introspection. Experimentation, physiology, and introspection are the methods of psychology. Alone they fill the world with fallacies; coöperating they give a valid psychology. Introspection has had the field to itself since the days of Aristotle and has filled the world with hallucinations. In these later days science comes with two new methods which, conjoined with the old, give promise of a new and better psychology.

In the compounding of judgments by sensation, if one consciousness is inferred to be like another then the present sense impression recalls that other. Thus the judgment of sensation is the judgment of likeness. A succession of judgments of this kind are consolidated in a concept and every additional sense judgment verifies the past sense judgment. When the present sense impression revives a past like sensation, it usually recalls it as integrated and differentiated. For example, I hear a sound and cognize it as a sound by recalling past sounds in a consolidated group, but in this case it may be a shrill cry. Another sensation of the same character may occur, the two being separated by a longer or a shorter interval; in this case I not only recognize the sound as such but also recall the former cry, so that I not only classify the cry among sounds, but also classify the cry among cries.

Every sense mechanically abstracts the impressions which it receives as distinguished from the impressions received from other senses. The eye abstracts sense impressions of light, the ear abstracts sense impressions of sound, the nose abstracts sense impressions of odor, the mouth abstracts sense impressions of taste, the skin abstracts sense impressions of touch, the muscles abstract sense impressions of force, as stress and strain, etc.

It does not comport with our present purpose to examine, either anatomically or physiologically, the nature of the senses themselves; we are simply trying to find out what a sensation is when we consider it as one of a group of like judgments forming a concept.

We see that the sensations are abstracted in that every sense organ recognizes a single property and that for every organ there is a fundamental property. Then we see that the sense impression coming into one organ is classified as like or unlike; thus the eye recognizes distinctions of light, the ear recognizes distinctions of sound, the nose recognizes distinctions of odor, the mouth recognizes distinctions of flavor, and the touch recognizes distinctions of texture. The muscular sense, or sense of strain, recognizes distinctions of force, and it is thus that sensation is abstraction and classification.

Kind is directly cognized by the sense of taste and odor. The same objects that are cognized by these senses may also be cognized by the other senses, and while they do not give direct deliverances of kind, they give deliverances which become symbols of kind. We cannot taste the kind when we touch the pear, but we can recollect it. We do not taste the pear when we weigh it in the hand, but we may recollect its taste. When standing under the pear-tree we hear the pear fall; we cannot taste it, but we may recollect its taste. When we see the pear upon the tree we do not taste it, but we may recollect its flavor. Thus the primary sense of kind is taste, and the other senses become vicarious senses of taste. We need a term for this faculty and shall use apperception to signify this cognition of different properties by one sense. Like all other terms of psychology, this one has been used in many senses with a tendency to universal meaning, but I shall use apperception to signify the union of judgments of disparate properties discovered by disparate senses. I have used concomitancy and comprehension to signify the union of disparate properties in one particle or body; in the same manner I use apperception to signify the union of judgments of disparate properties in one particle or body. This may be stated in another way. The development of taste is only the development of a cognition of an attribute, but all the five attributes or properties of bodies are concomitant, and though primarily recognized by disparate senses they are finally recognized as concomitants in bodies, and when a body is cognized by one sense it recognizes all of the properties of the body primarily discovered by the other senses. Thus in cognizing the property of a body by taste or smell, we may re-cognize the body itself with all its properties. In this manner one sense becomes vicarious for the others. This faculty we have called apperception.

We may consider a being so lowly that all its judgments are confined within the sphere of good or evil in the objects of the environment as they are related to itself as food. But if its fixed life were developed into a freely moving life, it would be guided in its search for food by an auxiliary sense of kind; this is the sense of smell. The primary sense is the sense of taste, but it has an auxiliary sense by which it discovers the same properties, for odors and flavors are the same, though gathered from the environment by disparate organs.

Verified judgments of sensation are cognitions of kind. Sense impressions of a kind are consolidated; this consolidation comes by experience and produces a concept; thus we have a concept of a particular color as distinguished from sound, or of sound as distinguished from strain, or of strain as distinguished from touch, or of touch as distinguished from taste. Sensation, therefore, produces concepts of kind, and the correlates of likeness and unlikeness are involved. We may define sensation as the cognition of properties as kinds in their effects, and it is a compound of judgments; and a judgment is a combination of a sense impression, a consciousness, a choice, a concept, and a comparison.

Such judgments as we have hitherto considered in this chapter are not the only judgments of kind which are formed by the mind. When a judgment is once formed and recorded in the structure of the brain, it may be recalled as a collateral suggestion of a sense impression, or by the will itself, and when thus recalled it may be compared with other concepts, and other new judgments of kind may thus be produced. The elements of a judgment of this kind are, first, the choice of a past concept; second, the consciousness of it; third, the choice of another concept; fourth, a consciousness of it; fifth, the comparison of one with the other. The products of these five factors will constitute a new judgment. Thus the constitution of the judgment still remains the same, but it begins with a recollection instead of with a sense impression. Thus judgments of kind are presentative and representative. Presentative judgments are inductive; representative judgments are deductive. By presentative judgments we accumulate facts; by representative judgments we generalize them under the law that whatever is true of an object is true of its serial or class identity.

An apple has the taste of an apple, the odor of an apple, the texture of an apple, the pressure of an apple, the sound of an apple when it falls on the ground, and the color of an apple when it is seen. Thus we have five methods of distinguishing an apple from a stone, a bush, or a bird. It will be noticed that I consider taste and smell not as disparate senses to distinguish disparate properties, but as varieties of one sense for the sake of distinguishing the same property. Thus we have five senses for discovering a body as a kind, and when a body is discovered as a kind by one of the senses this discovery may be verified by one or all of the other senses.

First we may verify a judgment of one sense impression by repeating the same impression, and finally we may verify what one sense impression successively affirms by an appeal to another sense. In deductive or representative reasoning the method of verification is at first by congruity of concepts, but when concepts are not congruous they may be referred back to presentative reasoning; this is experimentation. All generalizations or deductive conclusions may be referred back to experimentation for verification.

We may now give a more adequate definition of sensation. Sensation is a process of forming a judgment of number or of kind and of verifying the same. Verification is accomplished by repetition of the sense impression, or by referring the impression made on one sense to the court of another sense. In a case of judgment of number as distinguished from its correlate kind, man has devised a special method of verification known as measurement, which gives rise to the psychologic science of mathematics, which is also defined as the science of quantity. The judgment of number is verified by enumeration or counting.

We have found five classific properties: kind, form, force, causation, and conception, derived from the essentials by incorporation, and that the kind is a relative unit, the form a relative extension, the force a relative speed, the causation a relative persistence and the conception a relative consciousness.

There are no particles which are not found in bodies, and all bodies are composed of particles. The quantitative properties are found when we consider particles. Classific properties are found when we consider bodies. Thus quantitative properties and classific properties are reciprocals, and in each set there are five concomitants. The logician considers classes, the mathematician quantities; they thus view the universe from reciprocal sides; the one classifies, the other computes. Four of the categories are found in inanimate bodies, unless our hypothesis is valid. All five are certainly found in animate bodies. They all coexist and cannot be dissevered, so that when one is cognized the others are implied, and when they are all considered as kind they are subject to logical reasoning. In order that they may be subject to mathematical reasoning, kind must be resolved into number, form into space, force into motion, causation into time and concept into judgment, and then as properties they can all by substitution be represented by number, and thus computation is possible. It is only in the new science of psycho-physics that judgments are treated mathematically.

We may speak of a body without overtly affirming its properties, but they are implicitly affirmed or posited. The term posit is here used to mean the indirect assertion of something by directly asserting some other thing essential to it and in whose existence it is involved.

The word matter is the name of a collection of particles and every particle is a combination of essentials. The concept of matter has passed through the crucible of human experience and the most thorough and profound scientific investigation. All human knowledge, all scientific research, all ideation, and all logical expression are founded on this concept. To deny the reality of matter is to murder reason.

It may be well to recapitulate what has here been taught concerning substrates.

First, we have shown that the essentials of properties are their substrates severally; unity is the substrate of number, extension is the substrate of space, speed is the substrate of motion, consciousness is the substrate of judgment.

Second, we have shown that the quantitative properties are the substrates of the categoric properties; number is the substrate of kind, space is the substrate of form, motion is the substrate of force, time is the substrate of causation, and judgment is the substrate of conception.

Third, it has been shown that a particle and its essentials are one and the same thing, and that ultimate particles constitute the substrate of bodies. These self-evident propositions make the concept of substrate simple and clear.

The doctrine of bodies and properties herein expounded is simple. When it is compared with the metaphysical discussions of number, space, motion, time, and judgment, and the categories derived from them, which are kind, form, force, causation and conception, it will be a surprise to discover how tomes have been reduced to pages by eliminating fallacies. Censorious persons have sometimes accused the vender of beverages of adding water to wine. Brokers use this dilution of wine as a metaphor and speak of watered stock. It is astonishing how the vintage of science has been watered by the venders of speculation.

When the similar sense impressions come to an organ, relations of likeness are discovered; but when dissimilar sense impressions act upon the same sense their unlikeness appears. In this manner the sense impressions coming to the same organ are classified.

Then disparate sense impressions come to disparate organs, as light to the eye, taste to the mouth, etc. The same object may produce disparate sense impressions to disparate organs, so that at one time the object is a color, at another time it is a sound, at another it is an odor, at another it is a pressure, at another it is a touch, and at still another it is a taste. In this manner different manifestations of the same object are brought to the senses and integrated or unified as coming from one object, that is, the self learns that one object may have different manifestations; thus the apple exhibits color, sound, pressure, touch, taste, and odor. In this manner concepts are formed of different manifestations of the same body; thus sensation is the cognition of different properties in one body which is considered as a kind.

The self, having discovered the union of these manifestations in one body or particle, quickly learns that when one property is observed the others may be expected; thus the color becomes the symbol of the apple and it is known by sight, or the sound becomes the symbol of the apple and it is known by sound, the texture becomes the symbol of the apple and it is known by touch, the flavor becomes the symbol of the apple and it is known by taste, the odor becomes the symbol of the apple and it is known by smell. This is the recognition of an object by some one of its properties manifested to a sense and taken as the symbol of the object itself with its other manifestations and known as the cause of a sense impression. As the particle can be designated by naming any one of its essentials, so the body can be named by any one of its properties, and so also it can be recollected by any one of its properties. In perception a form becomes the nucleus of a concept which is recollected when a sense impression recalls it.

The lower animal, desiring to gather food for its offspring, and having the sense of touch as well as taste, could utilize its sense of touch in gathering food by the cognition of its form without resort to the sense of taste and yet it could verify touch by taste.

CHAPTER XV

PERCEPTION

It has been shown that there is a faculty of the mind, by which judgments and concepts of kind are produced, which has been called sensation. It is now proposed to demonstrate that there is a faculty of the mind by which judgments and concepts of form are produced which will be called perception. It is difficult to select a term for this purpose. It might be best to coin one, but the term perception seems to be more often used in this sense than in any other. There is a general sense in which it is used to denote all intellections, and there is a general sense in which it is used to designate all presentative judgments, but I use it to designate the making of judgments both presentative and representative, and also of concepts of form.

We must now set forth the process of perception as judgment. Here again we have a sense impression, a consciousness, a choice, a concept, and a comparison as the foundation of a judgment. The judgment or inference is that the two compared are caused by objects having the same or a similar form. In making the judgment there must be a discrimination and an identification. The mind having an object presented to it by a sense impression must choose some other concept of a form supposed to be like this form and compare the two and make a judgment of likeness or of unlikeness as the case may be. We thus see that the external form determines the internal judgment of form.

Like a judgment of sensation, a judgment of perception may be a certitude or a fallacy. If it is a fallacy it must be corrected, and if a certitude it must be verified. If the form were determined by consciousness there would be no need of verification, but as it is external, verification is necessary. A judgment of perception is imperfectly verified by repetition, for if the likeness is discovered a second time the judgment may be supposed valid, though the same conditions for error may still exist.

In perception, as in sensation, one judgment is certified by another of a disparate sense. If I taste and touch the apple I am sure that it is an apple, or if I taste, touch, and see the apple there is still further verification.

A sense impression of light falls on my eye and I infer that it was caused by a horse of which I had a previous concept. The inference or choice of this cause recalls this concept and I conclude that the impression and the memory consciousness are alike. The concept of the horse was the concept of a form; thus the cause was conceived as a form. Perception is cognition of the cause of a sense impression, considered as a form.

When we have recognized an object many times, the process of judging seems to be abbreviated by the cancellation of the act of choice; certain it is there is no conscious act of choice and apparently the judgment follows immediately upon the consciousness of the sense impression. This cancellation of some of the elements of a judgment is particularly observable with sense impressions of vision when introspection seems to reveal no intermediate elements. It is only in cases where original judgments are made and those where there is some obscurity in the sense impression, that all of the elements of the judgment are revealed. This phenomenon of apparent cancellation of elements of judgment that are made in the act of perception cannot too strongly be emphasized. Not only are elements frequently obscure or entirely lost, but whole groups of judgments seem to be canceled in the stream of thought.

That which has been called the choice, the guess, or the hypothesis, is not a random choice but is a choice which arises from experience.

I am wandering on the shore of the lake. Weary with a long walk, I climb to the summit of a rock, from which vantage ground I hope to obtain a better view while resting. In climbing I grasp the angle of a boulder over my head and immediately feel a pain thrilling through my nerves. From the sensation of touch I gather other knowledge, as I think that I have cut my finger on the sharp edge of a crystal. From where I stand I cannot see the crystal, but my knowledge of these rocks is such that I know that sharp crystals of feldspar sometimes protrude from them, and I think of it as such. My mind neglects the effect upon myself to discover its cause—a sharp crystal on the rock—and I have made a discovery. It is my present knowledge of boulders and crystals that guides me to this discovery. Without knowledge of this kind I might give some other interpretation to the sensation. If a moment ago I had seen a rattlesnake crawling over the grass,

I might have made a false interpretation and fancied myself wounded by the fang of a serpent. Or suppose I had seen a sweet-brier growing over the rock; then I might have concluded that my wound was from a thorn. This same sense impression, under different conditions of knowledge, may have different interpretations. The true interpretation is reached only because there already exists in my mind the related facts necessary to correct interpretation. The inference, therefore, is controlled by previous knowledge, and, in this case, guided to the truth.

Sitting upon the rock and gazing around the lake, my eye follows the meandering of the shore, and I readily distinguish the blue waters from the green banks. This perception is much like that by which the crystal was discovered. Let us see in what respect it is the same and in what respect it is different. In the one there was a sense impression of touch and a feeling impression of pain in my finger when the nerve was pricked, and in the other a sense impression on my eye when the nerves were touched by light, but no feeling of pain. The light reflected from the waters beats upon my eye and produces an effect, but I do not think of the sense impression as an effect, but only of its cause. The mind goes out beyond the consciousness to the object which produces it.

In the group of mental operations by which the crystal is recognized the particular feeling of pain is conspicuous; but in the operations by which the water is discovered, the beating of light does not cause a feeling as a pleasant or as an unpleasant effect. The discovery of blue waters and green banks cannot be made without previous knowledge. Suppose that I have never seen water—that I have suddenly been transported from some mythic land where basins of glass are embosomed in the landscape; with only such knowledge in my mind I think of a beautiful sheet of glass, and, though erroneous, the interpretation is believed as true, unless I submit it to verification.

Or suppose that I climb to the top of a mountain, where bays and inlets are thrust into the land. On arriving at the summit I look about, and the mountain seems to be an island. From the foot of the mountain on every side there seems to be a stretch of gray water. After a time a breeze starts up, and the water seems to be agitated in great waves, and at last the waves are driven away in tumultuous cloudlets. Now the blue lake stretches from the foot of the mountain on one side and valleys and hills from the other. My first inference was a fallacy; my second inference is a certitude.

I look along the shore again, and I see a white object on the water. What really happens is that arranged light reflected from the distant object beats upon the nerve of my eye, which differs from other light entering it. I do not stop to observe the effect on me, but my mind is occupied with the external cause. I am just from the seaside, and have been watching the gulls soaring through the air and gathering flotsam. I interpret the beating of this white light as caused by the reflection of light by a gull. I believe I see a gull; but it moves not, and I doubt the veracity of my vision. Looking again with care, I believe that the cause of this beating is a white boulder with its crest emerging from the water. Satisfied with this interpretation, my attention is directed to a boy coming down to the shore. As a sansculotte he wades into the water and follows the floats of a net until he comes to the white object which was to me first a bird and then a boulder. Now I make the true inference and see that the white object is a white cloth—a signal on the top of a stake to mark the fishing ground—and verify it. The facts uppermost in my mind caused me to make false interpretations, each of which I could not verify, and rested satisfied only when I made an inference that was verified. As perception by touch is the interpretation of a sense impression, so perception by sight is the interpretation of a sense impression. Here again we have an interpretation which gives a judgment of the external cause of the sense impression.

Still sitting on the rock, I hear a noise. It is but waves of air beating upon the nerves of my ear; but I go beyond the consciousness and turn my head in the supposed direction of the sound, expecting to see a man coming in the distance; for have I not heard his voice? At this I am disappointed; and yet it does not seem strange, for I have made erroneous interpretations many times. I continue to watch the fisherboy in the river below. The noise is heard again, and this time it is the caw of a raven in a distant tree. I have chosen the right cause. I muse on this error. Why is the voice of a crow mistaken for the voice of a man? Because I am expecting my friend who stopped by the way where blooming plants attracted his interest. A false interpretation of a consciousness often comes from expectancy. In this manner the deluded victims of the thaumaturgic seance are made to see and often to hear the very spirits of the dead and to find confirmation of fond belief. The human mind can discover any wonder the imagination can picture, however unreal or impossible it may be, if expectancy first be wrought to the requisite intensity. All perception is by interpretation, but the data by which we interpret are memories, and correct interpretation depends upon the right guessing in the first instance and ultimately on verification; once more we have verification necessary to cognition.

Inductive or presentative perception having been set forth, it is now required to explain deductive or representative perception. A concept of perceptive judgments may be brought into consciousness by an effort of the will or adventitiously by association, and this concept of form may be compared with other concepts of form and a representative judgment made about two concepts, both of which are recalled from memory.

Thus from the storehouse of memory we may take up by choice the innumerable concepts therein and make new judgments and combine them into concepts. These representative concepts can all be traced back to presentative elements.

A representative judgment of perception, like that of sensation, has five elements. Instead of the consciousness of the sense impression we have the consciousness of a concept brought about by an act of choice arising from association or exercise of the will. Then a second concept must be chosen or recollected, and then we must have a consciousness of this second concept, and when the one concept is compared with the other a judgment is formed.

In presentative judgments we compare an impression consciousness with a memory consciousness. In a representative judgment we compare a memory consciousness with another memory consciousness. In both cases we judge of likeness or unlikeness between the terms compared. In presentative judgments we discover facts and classify them; hence presentative judgments are inductive. In making representative judgments we discover laws and apply them; hence representative judgments are deductive.

We may now more adequately define perception as the process of making a judgment about form or its reciprocal space, and of verifying the same so as to produce a cognition. Verification is accomplished by repetition, by the same sense, by submitting the judgment to another sense, that is, by congruity of judgments or by submitting it to experimentation, which is also by congruity of judgments, or by submitting the judgment of form as its reciprocal space to measurement and computation, which is only another method of verification by congruity of concepts.

A strange confusion is found among some metaphysical writers in confounding the presentative judgment with image forming. Touch is the primal sense of form, but other senses perform the same task vicariously. As taste and odor are the senses by which we discover kind and the concept of form becomes the symbol of the kind, so on the other hand while touch gives us form the kind may become the symbol of the form. Now the sense of vision is highly adapted to the performance of this symbolic or vicarious function. The image which is cast upon the retina is but arranged color with an outline which is interpreted by vision to be the mark, sign, or symbol of a form, and the perceived image is a judgment. It is thus from vision that we derive symbolic judgments of form, and the judgment which we make is the image of the form. In vision the judgment of form is but one of the judgments we make; and there are as many kinds of judgments of form as there are organs of sense. Now the metaphysical doctrine which makes images and ideas to mean the same thing as presentative and representative judgments doubly confuses the subject, for thought is a succession of judgments of all kinds and image making is a presentative judgment of vision.

We more often make the form the symbol of the other properties of a body than any of its other properties. While form is primordially cognized by touch, and touch is the final arbiter in verification of judgments of form, yet vision is more facile in making such judgments and multitudes of judgments of form are made through the sense of vision where one is made by the sense of touch; notwithstanding this the judgments of vision are greatly subject to error and often require verification.

It is due to facile cognition and recognition of form by vision that the forms of bodies become symbols of all their properties. Bodies through their forms subserve many purposes, but they also subserve many purposes through their kinds and through the other properties which inhere in them, as forces, causes, and concepts. But we seem often to cognize them first as form. We see the forms of a thousand apples, peaches, or pears, where we taste but one, and so we habitually know apples, peaches, and pears by their forms; so we know all plants by their forms, but few by their tastes and odors; so also we know all animals by their forms and but few by their tastes and odors, though it would seem that the dog knows many more things by their odors; most rocks are known by their forms, few by their tastes; altogether bodies are known as forms much more than as kinds, forces, causes, and concepts, all of which is due to the fact that vision reveals form with such marvelous rapidity, while the medium of ether is unrecognized in making presentative judgments, and is discovered only through a long course of history in the development of representative judgments. It is not strange, therefore, that early metaphysical reasoning made such a profound distinction between impressions and ideas and confused judgments of form with images by reaching the conclusion that all presentative judgments are images pictured upon the retina. We paint images and the art is coetaneous with human culture. What we do by art in painting it was supposed that nature does in light upon the retina, and this is true within certain limitations, but the picture upon the retina must be judged like the picture upon the canvas, and in both cases the arranged colors are but symbols of form which is primarily learned by touch.

In forming deductive judgments of perception, that is, judgments of form, we may find that our concepts are incongruous, that one judgment contradicts the other. When this is the case one or the other must be erroneous; we are then thrown back upon experimentation for a verification of the past judgments of which these concepts are composed. Experimentation thus becomes the great agency in time for clarifying concepts and for purging them from error, that the inductive basis for deductive reasoning may be sound.

It has been seen how a stream of sense impressions pours into consciousness a body of symbols, which are there organized into systematic knowledge. Clouds assemble, change their hues and vanish; storms devastate the land and tempests vex the sea; the waters of the sea are lifted into the clouds, and the clouds themselves gather about the mountains and roll as river torrents in return to the sea; continents, islands, and mountains are upheaved, rains and rivers carve them into wonderful forms; volcanoes endeluge the land and trouble the sea; geologic formations are built and destroyed; the mountains, hills, plateaus, plains, and valleys are covered with the verdure of life; the air, the land, and the waters teem with animal forms; man himself is distributed over all land between the ice-formed walls of the polar regions—all the multitudinous objects of the cosmos are forever signalling to the human soul through vision and demanding its attention. Now one is seen, and now another; now one is heard, and now another; now one signals with fragrance, and now another; now one signals with flavor, and now another; and now one beckons with tactual signs, and now another; and the human soul gathers all these symbols into one gigantic body known as the human mind. The external world is thus coined into symbols, and of these symbols the foundations of mind are laid, and of these symbols the walls are constructed, and of these symbols the dome is reared, until the temple of the soul is perfected—a symbol structure built in every soul by the phenomena of the universe.

CHAPTER XVI

APPREHENSION

It has been shown that there is a faculty of the mind by which bodies are cognized as kinds, which has been called sensation. It has further been shown that there is a faculty of the cognition of bodies as forms, which has been called perception. It is now designed to demonstrate that there is a faculty of the mind by which bodies are cognized as forces, and this faculty I shall call apprehension.

A satisfactory term for this faculty is not found in the language. The term understanding has vaguely been used in this manner, but so many meanings for the term are in use that it cannot well be employed. The term apprehension also has several meanings, the most common of which is a synonym for fear, as when I affirm that I apprehend danger. I shall use the term apprehension as restricted solely to the judgment of force. Apprehension, then, is the name of the mental process of cognizing force in all its modes. In order that my argument may proceed I must have a term which will be taken with this meaning and with it alone. Whether I choose the term wisely or unwisely is another question.

Man is conscious of his own force, and he infers force of other bodies because of their effects when they impinge upon himself, being conscious of these effects. Then he discovers the forces of molar bodies in the change wrought by their impinging upon one another.

It has already been shown that man is primarily interested in the environment. The primitive man first becomes interested in what he supposes to be the environment of molar bodies by which he is surrounded. A vast multitude of these bodies are molar, and stellar bodies are at first supposed to be molar, while molecular bodies are unknown, and the world is supposed to be composed of molar bodies. Then human concepts are all of molar bodies and their properties.

In a judgment of apprehension, there are the same pentalogic elements that hitherto we have observed in judgments, namely, a consciousness of a sense impression, a choice of a concept, a consciousness of that concept, a comparison of one consciousness with the other, and a judgment which identifies or discriminates in affirming them to be alike or unlike as the case may be. The concept chosen is a concept of force. A judgment of apprehension must primordially follow a judgment of perception, just as a judgment of perception must primordially follow a judgment of sensation. This is the primordial order in which these judgments occur.

If we judge of external force in two bodies, before there can be a judgment of apprehension, there must be a plurality of judgments of perception as in perception there must be a plurality of judgments of sensation. When two bodies act upon each other a change occurs in both. In order that a judgment of their actions upon each other may be formed, there must be judgments of perception; the two bodies must be perceived. Then their action is inferred from the changes which they undergo; but it is impossible to have this judgment without the antecedent perceptions.

Let us consider a judgment of apprehension in what seems to be its simplest form. A pressure on self is experienced. Here there must be a sense impression which produces a consciousness, a discrimination, a choice, a recollection, and a consciousness of a concept out of which arises a judgment of simple sensation. Then we consider the cause as a form, and judgment of it is a perception. Then we consider it as a force in a process, and it is a judgment of apprehension. Thus a judgment of apprehension is one of a series of judgments, the first of sensation, the second of perception, and the third of apprehension.

We become expert in making judgments. Having made and verified them, cognition becomes recognition, and recognition seems to be a very simple process, for the pentalogic elements do not arise in the cortical consciousness. The fact is well known that judgments of intellection as well as judgments of action are made instantaneously with precision, when they have previously been made with halting labor, occupying much time; still we are compelled to the conclusion that judgments of apprehension can occur only after judgments of perception, and these only after judgments of sensation, although these several judgments all have pentalogic elements.

There seems to be in the mind or cortex a power by which logically antecedent judgments are cancelled after they have once been made, thus saving time and thought. This cancellation of the elements of judgment we have hitherto observed, and shall be reminded of it hereafter. This is the psychical phenomenon known as intuition.

Judgments of energy and work are primarily derived from muscular sensation or the sense of stress and strain. There is consciousness of stress when other bodies press upon us, and a consciousness of strain when our bodies press upon others. The consciousness is but a consciousness of change in self, but there is always an inference in a judgment. When I act I am conscious of the action as a cause, and infer the effect; when another acts upon me I am conscious of the effect and infer the cause. But here we do not pause to treat of the consciousness of strain, but only the consciousness of stress.

The faculties of intellection, which we have called sensation, perception, and apprehension, are connate; that is, they are contemporaneous growths as concepts, but not contemporaneous judgments. The judgments of sensation must precede the judgments of perception, and these precede the judgments of apprehension. The last judgment formed may seem to follow upon the sense impression itself. It is the power which seems magical to the untrained psychologist; the power of reaching a conclusion by previously gained knowledge. It is the power which we call habit in another realm of psychology, as when the trained pianist strikes many notes simultaneously in rapid succession. Here we observe that the successions of the mind are more rapid than the fingers, for the successive acts of will for every finger are interpreted by simultaneous muscular acts. It is the power by which the intellect considers many judgments in such rapid succession that they appear to be simultaneous.

Heretofore, in discussing sensation, perception, and apprehension, the effect has been subjective and the cause objective, but in apprehension these relations of cause and effect are sometimes reversed, and the cause may be subjective and the effect objective. I am conscious not only when another strikes me, but I am conscious when I strike another. Here we have a consciousness of cause, and the effect is inferred. I am conscious of a flavor when I eat an apple, and I am conscious of an act performed by myself when I bite it. I was conscious of an effect of color when I saw it, and I was conscious of an effect upon myself when I touched it; I was conscious of an act when I turned my eye to it, and I was conscious of an act when I grasped it. Thus there is always an emotion connected with an intellection and there is always an intellection with an emotion. But we are not now considering emotions; we are considering intellections only. We cannot consider intellection without positing emotion. With this statement we go on to consider the subject of the intellections, our present purpose being simply to discover an epistomology for the intellections. In another book we shall treat of the epistomology of the emotions.

Here again we must call attention to another very important fact, viz., that the individual mind is only one of many minds, and that it is only one of a still greater number of bodies—that there is myself and the environment, and that there is yourself and the environment, and that you are a part of my environment, and that I am a part of your environment. Thus every body in turn is a self with an environment. The wind acts on me, and I act on the wind, the tree acts on me, and I act on the tree; but the wind acts on the tree and the tree on the wind, and these actions are all processes. The action of the wind on the tree and the tree on the wind come into my judgment, and these processes are cognized; but the cognition of the action of the wind on the tree and that of the tree on the wind is inferred by their actions severally upon me. I see the leaves on the tree stir in the wind, but I am not conscious that the wind stirs the leaves. I am conscious that the light from the tree impinges upon my eye, and *infer* the tree and the wind with all the processes involved. Thus it is that the cognition of action and reaction between objects in the environment is a very complex process of reasoning, for cognition of the interactions of the objects of the environment are composed of a vast congeries of judgments.

Force must not be confounded with causation, although there can be no causation without force, nor can there be force without form, nor can there be form without kind; but abstractly causation and force are wholly disparate. A sledge impinges on a tree; the sledge strikes the tree and the tree strikes the sledge; action and reaction are equal, and in both vibrations are set up which are visible in the tree but invisible in the sledge, though none the less real. Now, when I consider action and reaction, I am considering force; but the sledge makes a visible indentation on the tree. When I am considering this indentation, I am considering an effect; perhaps the sledge changes some of the relations of its particles in crystallization; when I consider this effect upon the sledge I am considering causation. Suppose that, instead of striking the tree with a sledge, I strike it with an ax, then the blow produces a cut; and when I consider the difference between a cut of a sharp ax and the indentation of a sledge, I am compelled to consider differences of causation, and though the force of the blows are equal, the forms of the cause are unequal. The blow on the tree causes both vibration and indentation. Thus there are two effects, but only one blow. The blow on the sledge is vibration and crystallization; but there are two effects, but only one blow. When we consider the nature of the blow as action and reaction, we are considering force; but when we consider the effect we are considering causation. Action and reaction are simultaneous, cause and effect are sequent.

All intellection is abstraction; he who cannot accomplish and hold firmly an abstraction cannot psychologize.

Apprehension is both presentative and representative, or inductive and deductive. If we look upon apprehension from the standpoint of its initial element, it is either presentative or representative; but if we look upon it from the standpoint of result as reason, it is inductive or deductive. The choice of a concept of deduction is always initiated by choice of another concept instead of a sense impression. This choice of a concept may be the one made in a presentative or other judgment, for judgments may follow judgments in extended succession, all initiated by one sense impression, but connected in the succession by links of recollection. From one point of view these may be called discursive judgments, and from another associated judgments. In waking hours the mind cannot cease to make judgments. If sense impressions are neglected, recollected concepts take their place. The mind may be turned loose to make excursions by steps of judgments into a field where fancy leads; but the path of the mind in making judgments may be directed by the will to the accomplishment of a purpose, in which case the judgments instead of being discursive are volitional. Representative judgments, therefore, are discursive or volitional.

I see a bird flit from one bough to another. If my mind is free to pursue my meditations, I may recall the bird that I saw yesterday, and this may recall a nest of blue eggs, and this may recall the blue scarf of my little daughter, and I may go on in this manner to make discursive judgments; but I may be watching the movements of the bird for the purpose of studying its habits, and my judgments may be controlled by my will. In experience we pass from presentative to representative judgments, back and forth, with instantaneous rapidity and great irregularity. So we pass from discursive to volitional judgments instantaneously and irregularly.

Judgments become cognitions only when they are verified. Judgments of sensation are verified by submitting them to other senses, and then they are subjected to perception for further arbitrament. Judgments of perception are submitted to apprehension for verification, but judgments of apprehension are verified by a faculty which we have hitherto not discussed. We must now set forth the office of apprehension in verifying judgments of perception.

Forms are not properly conceived until we know

their function. We may have a vague concept of a form without knowing its function, but the elements of its structure are not fully grasped until we discover their relations to function. Thus our perceptions of form are not only verified by our apprehensions of function, but the observation by which it is discovered is often dependent upon the effort to apprehend function. An obscure stigma on the pistil of a plant might be wholly unobserved by the man who is not acquainted with the office of the pistil, but the botanist is sure to perceive it. The painter perceives muscles with certainty when he observes them in action. It is thus that perception is verified by apprehension.

In the human race, knowledge commences by the cognition of molar bodies; as culture advances knowledge is extended to stellar bodies in the direction of the vast, and to molecular bodies in the direction of the minute. On the other hand, knowledge has not only been extended into the vast and the minute, but it has also been extended into the compound and complex as exhibited in plants and animals. This distinction has long been recognized in a vague way by including certain sciences under the term natural history, and other sciences under the term physics. The real distinction between these sciences, however, is this: that the natural history sciences consider quantities or properties that can be measured. In ethronomy and astronomy we consider properties that can be measured, and ultimately arrive at classification; but in phytonomy and zoönomy we first consider properties that can be classified, and finally resort to their measurement. In geonomy the sciences are broadly grouped into two classes, namely, geography and geology; the geographic sciences are sciences of measurement, the geologic sciences are sciences of classification. Thus we have quantitative and classific sciences. This is the old distinction of metaphysics between quantitative and qualitative things when properties are considered as qualities. We have already seen in the chapter on qualities the nature of this error and are ready to rescue the term quality from the ambiguity into which it fell when it was considered as synonymous with class, kind, or category.

The so-called qualitative sciences, therefore, are more properly designated as the classific sciences. This broad distinction between the classific and the quantitative sciences deserves some further consideration. In the deductive sciences there must be some reason why we first look for quantity—why we come to study the ether, the stars, and geography quantitatively, and geology, plants, and animals classifically or categorically. We know absolutely nothing of kinds of ether, but only of the properties of ether as belonging to one kind. We know of no method by which we can change the particles of ether into kinds. We know of but few kinds of stars, and we know of no method by which we can change the kinds of stars. There are but few kinds of air and of water, and these differences are only varietal, not specific, and the elements of mathematical geography are established mainly beyond the interference of man. We wish to adjust our conduct to these established facts, and hence we wish to know the facts. I do not propose to change the rising and setting of the sun, but I do wish to measure the times when they may be expected and the length of the day and the night. I do not propose to change the gravity inherent between the several stars of the solar system, but I do wish to measure the force of gravity between star and star, that I may adjust my conduct to established facts when I make the ephemeris for the guidance of the navigator. I do not propose to change the atmosphere, but I measure it by determining its barometric quantities, the pressure of its winds, and the quantity of moisture which it contains. In the same way I measure the superficial extent of the sea and the depths at which the rocks are found, that I may adjust my conduct while navigating the sea to the facts therein discovered. Now, we could go on to illustrate these facts in a multitude of ways, and in an endless procession, and find in all those realms of science, which I have indicated by calling them physical or quantitative, that I am interested in quantities as a dweller upon the earth.

In the quantitative sciences there are few kinds, but many of a kind. Induction is the discovery of a kind; deduction is the application of the laws of a kind to the individuals which are included in the kind. The quantitative sciences are deductive, for deduction predominates in their study. It is thus that the physical sciences, ethronomy, astronomy, and geography are quantitative and deductive, and that which interests us most in these realms of bodies is their quantities, for though it is impossible to change their kinds, it is possible to adjust ourselves advantageously to their quantities.

In geology, phytonomy, and zoönomy there are many kinds; thus there are many mineral species and few individuals of a species as compared with the individuals of air and water; there are also many kinds of plants and comparatively few of a kind. Deduction is based upon the law that what is true of one of a kind is true of all of a kind, but where there are many kinds and few individuals, attention must be given more to the discovery of the kinds than to the application of the laws to individuals; hence in these sciences our attention must relatively be occupied with the discovery of kinds and less occupied with the application of laws. But more: Man by culture undertakes to change kinds, forms, forces, causes, and concepts. By the arts of constructive or synthetic chemistry and metallurgy, he makes many new kinds. By a great variety of arts he makes many new forms, shaping the rocks, plants, and animal substances into a variety of tools, utensils, machines, and fabrics. In a vast multitude of ways he seeks to change the forms of bodies which he discovers in rocks, plants, and animals, and can accomplish his purpose only by changing the kinds.

So man tries to change the forces of nature into modes which he can control, and all of these changes which he brings about upon the face of nature depend upon his recognition of causes and the wisdom of his selection of the nature of the cause; while he is thus employed in changing the kinds of things in nature, he is forever building up and changing his concepts, and all of this change when resolved to its simplest statement is change of kinds. Thus in the geologic, phytonomic, and zoönomic realms, man is primarily interested in kinds and only secondarily in quantities; in the products of his cultural activities he is equally interested in kinds and quantities. It is here that induction and deduction meet on equal grounds, for the arts are equally inductive and deductive.

Presentative reasoning is thus chiefly classific and inductive, while representative reasoning is chiefly quantitative and deductive.

Now we see why bodies are symbolized as kinds rather than as forces, for forces are recognized as processes. We think of the force of a form rather than of the form of a force. All of the senses are under the control of and associated with muscles, so that we cannot taste an object without employing the muscles of the mouth, and when we designedly smell an object we must imbibe its vapor in the air through the action of our muscles by inhalation. We cannot touch an object without employing our muscles by extending the organs of locomotion, as hands or feet, though the object may touch us independently of our self-activity. So pressure is apprehended by us as stress or strain; the muscles of the ear are strained when we intently listen; the eye is especially under the control of a system of muscles, so that it becomes the special organ for the cognition of motion. We do not see motion chiefly because of the passing of the image across the retina, but because the eye, through its muscular apparatus, adjusts the point of vision of the image upon the retina to the moving body. Thus, while force is primordially cognized by the muscular sense, even motion comes to be cognized by the muscular sense when it adjusts the organ of vision to motion. This is one of the characteristics of vision which eminently adapts the sense to vicarious faculties.

We may now give a more adequate definition of apprehension. Apprehension is the process of forming a judgment about force, or its reciprocal, motion, and of verifying it so as to produce a cognition. The difference between a cognition of motion and of force inheres mainly in the method of verification. The various methods of verification are fundamentally dependent upon congruity of concepts. Again, apprehension as a process of intellection may be defined as the cognition of force.

CHAPTER XVII

REFLECTION

We have now to describe that faculty of the intellect by which concepts of causation are produced. It will be remembered that the essential, constant, or absolute of this property is persistence, that the relative is change, and from the two time is derived; then, as motion becomes force through the collision of particles, time becomes causation as antecedent and consequent, or cause and effect; then, causation becomes metagenesis, and metagenesis becomes heredity, and heredity becomes evolution.

Words are used with many meanings, but in science we are compelled to use them with one meaning. All psychological words are singularly ambiguous, because they are used as tropes to such an extent as to conceal their fundamental meaning. It is necessary to select a word to signify the cognition of causation, or cause and effect, in the various phases of time and evolution, and I select the term reflection for this purpose. The term may also have a meaning synonymous with contemplation, but I select it with the meaning which is involved in it as a sign for the cognition of causation.

Once more it may be well to remind the reader of the total unlikeness of the properties of matter, so that they can not be classified. Things can be classified that are partly alike and partly unlike, but properties are totally unlike. We may consider properties separately, but this is abstraction, not classification, and we may schematize the properties. Fundamentally, we reason by abstraction because we consider properties severally. By reason of the total unlikeness of disparate properties, the most fundamental and clearest distinctions in psychology are those which we make when we call a faculty the cognition of a property. Reflection is one of those faculties because, as the term is here defined, it is the cognition of the property of causation.

Reflection, also, has the pentalogic elements, but in the inference the choice is of a concept of causation. These pentalogic elements are a consciousness of a sense impression, a choice, a concept, a comparison, and the judgment of likeness or of unlikeness.

Reflection is one of a series of judgments, and by its place in the series others are presupposed or posited. The series, so far as it has been built up, is composed of sensation, perception, apprehension, and reflection. I see an oak, and may make a judgment of sensation and conclude that it is green. I see an oak, and I may make a judgment of perception and conclude that it is a tree. I see an oak, and may make a judgment of apprehension, and conclude that its leaves and branches are in motion; I see an oak, and make a judgment of reflection and conclude that the motion in the tree is caused by the wind. These judgments differ from one another in the nature of the concept recalled, and these concepts differ in degrees of compounding.

Why do I make a judgment of sensation? Because I wish to note the color which I am painting. Why do I make a judgment of perception? Because I wish to seek the shade of the tree? Why do I make a judgment of apprehension? Because I am looking for birds. Why do I make a judgment of reflection? Because I wish to note the direction of the wind. Here again we see that the particular inference which we make depends upon the choice of a concept, and that this choice of the concept depends upon our purpose.

The concepts of reflection are compounded of judgments of causes and effects of events. Thus by reflection the relations of time are compounded into the relations of causation, and then these are compounded into relations of metagenesis, and these are compounded into relations of heredity, and these are compounded into relations of development, and these are compounded into relations of evolution.

It will be seen that the concepts of causation are exceedingly compound. In the practical affairs of life, events are of profound importance, for the events of yesterday affect the events of today, and those of today will have a consequence in the events of tomorrow; thus life is a constant discipline.

The time of which we speak is not void time, but the time of states and events, for of void time we know absolutely nothing, and language fails to express any concept of void time, and any reification of it is a pseudo-idea—a mythological notion.

It must be understood that, as the cognition of form comes by experience, so cognition of force comes by experience. Cognition of form antedates the cognition of energy only in the sense that the full knowledge of form is necessary before there is full knowledge of force; the experience upon which they both depend is contemporaneous. This may be stated in another way to be made clear. Cognition of kind by sensation arises with a certain degree of experience; cognition of form arises with a higher degree of experience; cognition of force arises with a still higher degree of experience; but judgments of kind, judgments of form, and judgments of force are accumulated contemporaneously. So concepts of causation succeed concepts of force; but the judgments of causation are contemporaneous with the judgments of force, form, and kind, and there can be no judgments of causation without judgments of force, form, and kind.

Here we arrive at a paradox, as it seems, to those who fail to comprehend the nature of causation. Consider a valley down which a river runs. There can be no river without a valley, yet the river has caused the valley. You affirm that the river has carved the valley, which seems to be a paradox; there must have been a valley in order that the water should be gathered into a stream; and that the river presupposes or posits the valley.

You explain that a small tract of land is gradually left bare by the retiring sea, that is, the land is slightly upheaved; the rain falls upon the land and carves channels, the tract of land is extended, new channels are formed and the old channels are deepened; still the upheaval goes on with increasing dry land, multiplication of channels, deepening of channels, and the widening of channels into valleys, and this continues until at last a great area of land is upheaved from the sea, and the rains have carved channels and the channels have coalesced again and again until a great valley is formed through which a river rolls. The river in the process of its growth has carved a valley, and the enlarging land has at last caught water enough to fill a river; the growth of the valley and of the river are contemporaneous, but the forming of the valley logically succeeds to the falling of the rain and the flowing of the river with its lateral streams; that is, effect succeeds cause.

This is the metaphysical fallacy which mistakes an effect for a primordial cause, and practically says that the valley existed before the river, for it gathers the rain which constitutes the river. The valley was from the first, but the river is caught by the valley from the rain which falls. Examine the doctrine of presupposition in metaphysics, and in every case a fallacy will be found.

In this manner the experiences of sensation, perception, and apprehension are connate, they spring up together, and yet concepts of sensation precede concepts of understanding, and concepts of apprehension precede concepts of reflection. One part of the doctrine of presupposition, as it is put in metaphysics, is a fallacy, and is replaced by the doctrine of causation, which explains that that which was supposed to be antecedent is consequent, or that which was supposed to be cause is effect. This is the great contribution made by science in demonstrating the laws of evolution. Another part of the metaphysical doctrine is erroneous in assuming that the concomitants or properties are derived one from another, one school affirming that all of the properties are derived from force, the other that they are all derived from intellection.

There is a valid concept involved in the use of the term presupposition, so often occurring in metaphysics, for when one property is considered abstractly the others are known to exist; though not overtly affirmed, they are implied, and presupposition used in this manner and understood in this manner would be just as good a term as implication or concomitancy; the term presupposition leads astray when it suggests the further idea that the things implied are antecedent things, instead of antecedently known things.

Judgments of evolution are constituted in the same manner as other judgments, and to become certitudes they must also be verified. But judgments are consolidated as habits of thought; thus we come across the phenomena of intuition. When the mind makes one judgment and uses other knowledge which was derived by previous judgment to make a new judgment apparently far remote from the first, this new judgment is said to be a judgment of intuition, for the steps seem to be cancelled in reflection, and the long course of reasoning is made to appear as a direct result.

I see the track of a man in the sand. The left track is full, the right track shows only the impression of the toe. I see the one and then the other, and I infer that the man was lame and walked upon his right toe. John Smith is lame, and I infer that John Smith has walked along the trail. John Smith lives at a distance; I have heard that his mother is ill, and that he has been sent for, and I infer that he has passed along the trail to the home of his mother. Thus a series of judgments flash through my mind when I see the half footprint, and so speedily do these judgments arise in succession that the intervening steps seem to be cancelled from intellection, and I appear to infer from the footprint directly to the visit of John Smith to his mother; but in fact I have carried on a series of judgments derived from elements of knowledge that have been recalled by the sight of the footprint.

This reasoning in series by unrecognized steps is intuition. It is the same old story of habit. Certain kinds of reasoning, like certain kinds of muscular activity, come by frequent repetition to be so easily accomplished that the processes involved are unrecognized by the mind. Perhaps this can be explained by the theory that in recalling one concept we recall others with which it is associated, reviving them as they are woven into the structure of the cortex by the act of choice. All judgments of causation are more or less serial in this manner, and as most of them are habitual they become intuitive. For this reason it is often more difficult to analyze judgments of reflection than judgments of apprehension; and more difficult to analyze judgments of apprehension than judgments of perception; but by careful attention to the subject and by the acquisition of skill in introspection, it can always be discovered that every judgment of reflection is founded upon a consciousness and involves an inference which recalls a compound concept, and to reach the stage of certitude it must be verified. Finally, it must be remembered that intuition, which is supposed by careless thinkers to be occult, is in fact developed by experience. Such is the nature of presentative judgments of reflection.

We have yet to consider representative judgments of reflection. Again, we see that as presentative judgments follow upon sense impression, so representative judgments follow upon choice, and the choice may be discursive or volitional. The discursive choice is sporadic, and by following such concepts the stream of thought is directed in a meandering course that flows to nowhere; but the choice for a fixed purpose, in which there is an interest, leads to results that influence the conduct of life. The presentative judgments of reflection are removed from the sense impression by intuitional or by more deliberate judgments of sensation, perception, and understanding, so that the judgments of reflection, both presentative and representative, are more deliberative than of the lower faculties.

When both cause and effect are external, the judgments of them are mediated by other judgments, the causes of which are external and the effects internal; hence the judgments of external cause and effect are still further removed from sense impression, so that there is again another degree of deliberation. It is this characteristic that has led to the selection of the term reflection to designate the faculty, and although the reflective judgment may never have been defined as it has been here, yet this definition will serve to reveal the unconscious wisdom of the selection of the term in current speech. In judgments of original cognition the pentalogic elements can always be discovered by introspection, but in the judgments of recognition it is difficult to discover them in the cortical consciousness. When cognition is fairly accomplished recognition thereafter becomes instantaneous.

Audition is the primordial sense of causation. Sound comes to us through a medium, and primordial man has no knowledge of this medium; he does not recognize the ambient air. Thus he thinks that sound is something emitted from bodies, just as Newton believed that light was something emitted from bodies, and Plato that forms were emitted from bodies. So the savage looked about him for the cause, and often the cause as a form he could not see, and as he knew nothing of molecular force he formed no concepts of force in relation to sound; so his concepts of sound were concepts of cause until he could discover the cause as a form. It was thus that concepts of cause were primitively generated in the mind of man. Hearing is also the sense by which time, the reciprocal of cause, is first conceived. We must remember that properties are concomitant, and though the faculties operate abstractly in that they primarily conceive properties as abstract, yet the indissolubility of the concomitants compels us to consider the manifestation of one property as the symbol of all others. In this manner the senses all become vicarious, and we make judgments with one sense that we might make with another. Of all the primordial senses we have hitherto discussed as the primal sense of a faculty, that of hearing is the most facile to perform the functions of the others, though we shall hereafter observe that seeing is the grand vicar of the senses.

Judgments of reflection are verified by the judgments of a higher faculty, but they themselves are used to verify the judgments of lower faculties. Motion and force are expressed in rapidly passing events, but causes produce effects that remain; causes and effects are states; forces are events that separate states; hence it is that the judgments of understanding are relegated to those of reflection for verification.

I suppose that I see a woodpecker tapping a tree. I look and see the fresh pit made, and my judgment is confirmed. I obtain the glimpse of an animal running through the forest, and think it to be a wolf. I come to the spot where it was supposed to be, and the tracks of a deer are seen, and so I correct my judgment. Thus a higher judgment will serve as a verification of a lower.

Judgments may be measured. I judge of a distance, and find, when the distance is measured, the error of the judgment. I do not find the error of the line measured, but only the error of the judgment made. So, whenever we make a judgment of length or distance or size or weight or mass, or what not, we measure our judgments by measuring the what-nots judged. All judgments are liable to error, and cognition comes only with verification. In quantitative judgments the liability to error is infinite as that term is used by mathematicians, and all judgments must be verified unless the amount of error may be neglected.

In scientific research verification is often by measurement. Counting itself is measuring, and the sum is the number of units which the measured body contains, and these units are units of a kind. It is only in counting that the units are natural; all other units are conventional in that something other than the thing measured is taken as the unit or standard. I measure time by the revolution of the earth, by the revolution of a hand on the dial of a clock, or by the flow of water from a clepsydra. Thus one measurement is mediated by another, and different standards are taken. The nature of measurement is well understood except in so far as it relates to psychological phenomena; in this realm metaphysicians seem wholly to misconceive its nature. I cannot measure the number of the ultimate particles of a body by counting them, but I may measure the relative number of its atoms by weighing it. I do not determine its force, but its mass only, when I weigh the body, for the total force in the body is the sum of its motion in all its incorporations. A pound of powder has much more force than that which is measured as a pound. What we really arrive at in weighing a body is the proportionate number of its particles. I may measure the length of the wall by counting the brick lengths in the wall. I cannot measure this stick by counting the number of particles as atoms, molecules, or cells which constitute its length, but I use a conventional unit, say an inch, and I find it ten inches long. Had I taken some natural unit I might have found it, say, ten million molecules in length. Now, what have I measured? Only the distance which separates the positions of the molecules in its termini, but I have not measured the extension of any of the molecules, for probably they are separated by interspaces filled with ether, and may be with air. It is thus that I measure space. I cannot measure form, for form is internal structure and external shape.

I have a body which is of very irregular shape, and hence I cannot well determine its extension in three dimensions, but I put it into a beaker of water, and determine how much it displaces, and measure that; thus, while I do not measure the body itself, I measure its equivalent. Now this leads us logically to the statement that of the five concomitants in every particle or body every one can be measured, and it is only necessary to measure one property to have a measure of them all. But more than this, I must measure one property in terms of another; thus, I measure motion in terms of space or length, and I measure speed in terms of length and time. We must remember that measurement is always a conventional process to serve a purpose, and the way in which we measure a thing is by some device for the purpose, and the purpose is always the relation of the thing measured to some other thing. I measure force as motion, so I measure the force of the cause in its effect, and measure the effect in space elements. I measure a judgment by measuring the thing of which the judgment is made; thus, I judge of a distance, and may measure the distance to determine the amount of error in my judgment. It is in this manner that I can measure judgments.

I cannot longer dwell on this subject to set forth the devices by which judgments are measured, but must content myself with the statement that the attempt to measure judgments has but recently been made, and that already there are many devices. All of the properties can be reduced to or considered as number. Space can be considered as number when its elements are counted in natural units, or it can be considered as number when its elements are measured in conventional units. Motion can be considered as space, and then as number. Time may also be considered as motion, then as space, and finally as number;

judgments may be considered as time, and time as motion, and motion as space, and space as number. The device by which the other properties are considered as number is measurement, and measurement is experimentation.

We are prepared to give a more adequate definition of reflection. Reflection is the faculty of cognizing causation. Again, we may define it as the process of making a judgment about causation or its reciprocal, time, which judgment must be verified to become a cognition.

CHAPTER XVIII

IDEATION

We have seen that consciousness is one of the essentials of an animate particle; that sensation is the first mental property or faculty of an animate body; that perception is the second mental property or faculty of an animate body; that apprehension is the third mental property or faculty of an animate body; and that reflection is the fourth mental property or faculty of an animate body. Now, we have to consider the fifth mental property or faculty of an animate body. We form judgments about consciousness and choice, and about judgments and concepts; that is, we cognize mind. We need a term to express the forming of judgments about judgments, or of cognizing cognitions. For this purpose I shall use the term ideation. Ideation, therefore, as the term is here used, is the act of making judgments about judgments which, when verified, are cognitions.

We are conscious of our own judgments, but we infer the judgments of others. We may find the judgments of others to be like those we have already formed, or we may find that they are new to us. These new judgments we may accept or reject. When speech is developed and education is instituted, acception comes to play a very important rôle in mental acquisition.

Judgments of ideation are connate with all other judgments, but they are compounded of them and represent higher degrees of relativity; hence it is more difficult to trace them into their constituent judgments, yet trained introspection accomplishes this feat.

Before the laws of evolution were discovered and an absolute difference between man and the lower animal was supposed to exist, it was often affirmed that this distinction consists in the absence in the brute of knowledge about mind, that only man knows himself to be a thinking being, or, as we are here using the term, only man has the faculty of ideation. This is one of the affirmations which men are ready and prone to make before they learn that cognition is verified judgment, and that our judgments are guesses, while guesses are often more current than certitudes. With this idea was associated another, namely, that animals do not reason, but have instinct, there being no realization of the fact that certain practical judgments are repeated so often that they become intuitive as acts become habitual. Instinct or intuition and habit will require further consideration in a subsequent book.

We must now develop a little further the nature of the faculty of ideation, by considering the process of forming judgments of ideation. I hear a voice, and by experience know that its tone expresses surprise. Thus I form a judgment of an emotion in another. I am confronted with an antagonist on a field of battle, and see him point his howitzer at the column of troops in which I move, and infer that he has a deadly purpose. The lower animal makes judgments of ideation in this manner, and uses these judgments in guiding its own conduct. With mankind in higher culture this faculty is greatly developed. All words are signs of concepts, and all combined words that express thought are judgments, and the symbols of ideas, both spoken and written, constitute the pabulum of higher culture. Thus we not only cognize the intellections of others, but at the same time we accept their judgments as judgments of our own.

Ideation, as the term is here used, is a cognition of intellections in that manifestations of intellections are cognized by forming concepts of them. In sensation manifestations are conceived as expressions of kind. In perception they are conceived as expressions of form; in apprehension they are conceived as expressions of force; in reflection they are conceived as expressions of cause and effect; in ideation they are conceived as expressions of mind.

The constitution of the judgment which has already been exhibited four times must here be repeated. It appears as a consciousness of a sense impression, a choice, a reproduction of a consciousness of a concept of ideation, which, by comparison, make a judgment of ideation. The concept which is reproduced by the choice, is still more highly compound than in the lower grades of cognition, for the acts which animate bodies perform are first interpreted as kinds, then as forms, then as forces, then as causations, and finally as concepts. The series is complete when the judgment of ideation is made.

Like all other judgments, those of ideation are presentative and representative; representative judgments are discursive and volitional. There is no need to repeat the discussion setting forth the nature of judgments in these respects.

Vision is the primordial sense of ideation. We

see the motions in others, which I have heretofore called self-activity, and interpret them as symbols of soul. By soul I mean all intellectual and emotional judgments made by the animate being. The individual is conscious of the judgments made by himself, but he infers the judgment made by others. The judgments that others make are inferred from the signs which others make. I see the leaves tremble, the clouds move, the rain fall, the river flow, and innumerable motions in the mineral world; but I do not consider them as signs of intellect and emotion. There are other signs, however, which I observe in animate beings, and especially in human activities, which I do interpret as marks of soul. These signs are those which are produced only by those bodies which, being animate, have motility. The nature of this motility we have elsewhere explained and we have called it self-activity, which must not be confounded with self-motion, for self-motion is inherent in every particle, while self-activity is self-directed motion in a body.

Only animate bodies have this self-activity. But according to our hypothesis the ultimate particles of inanimate bodies have self-activity in so far as they manifest choice or affinity, while plant bodies seem to have self-activity in their cells. Neglecting this hypothesis, animate bodies certainly have consciousness and choice in their cells. Now, as one inanimate body has inherent motion in its several particles, which are organized in a hierarchy of bodies, the inanimate body cannot be deflected except by collision with another body, but the animate body can deflect its own motion as a body by metabolism, and by deflecting its own motion as a body it can deflect the motion of others. It is this power in the animate body of deflecting its own motion at will and of deflecting the motions of others by colliding against them at will, which is the sign or mark of mind in those bodies to which we attribute mind, and which exhibit more and more the purpose and ability to convey concepts to others, until among the higher animals a conventional sign language is produced, which becomes oral in the higher animals, but oral and written speech in man. Without words only emotions can be conveyed, whereas with words intellection can be exchanged. Gesture language may become gesture speech, oral language oral speech, and picture writing written speech. It is with this higher condition of language as speech that we are chiefly interested in ideation. Everything in nature has manifestations which may be interpreted, but only animate beings purposely convey concepts to one another.

Ideation is reënforced by other demotic agencies than those of speech. The pleasures, the industries, the institutions, and the opinions of mankind, are all expressed as human activities, and manifest the concepts by which they are produced; but we need not dwell on the subject here.

Through the agency of language we discover the fifth property of bodies. When we are interested in them and interest grows apace we may wish to know what those bodies say instead of what they are; it is then that language becomes speech, but culture continues to advance and speech becomes designed or purposeful instruction. Then all the appliances of instruction are developed until one of the principal occupations of mankind is the giving and receiving of instruction and the acquiring of concepts from one another, in which process the instructor is more instructed than the pupil, for the speaker in the organization of that which is spoken learns more than the hearer.

Now the eye, by its peculiar construction with apparatus for accommodation to distance and direction, is especially adapted to the reception of sense impressions that imply self-activity, hence it is the primary sense organ for the faculty of ideation. While its fundamental function is ideation, by reason of the concomitance of properties it becomes a vicarious organ for others.

Every one of the sense organs becomes an organ for and of the faculties. In the first stage of mind, while the organs of taste and smell are primarily the organs of sensation, the other organs interpret the sense impressions coming to them as symbols of flavor. In the second stage, while touch is the primary organ of form, the sense impressions coming to the other organs are interpreted as symbols of form. In the third stage the muscular sense is the primary organ of understanding, but all the other organs interpret the sense impression coming to them as symbols of forces. In the fourth stage, while the organ of audition is the primary organ of reflection, all the other organs interpret the sense impressions coming to them as symbols of causation. In the fifth stage, while the eye is the primary organ of ideation, all the other organs may interpret the sense impressions coming to them as if they were symbols of concepts.

We have seen how the judgments of the lower faculties are verified by the higher, but now ideation is the court of last resort. In the structure of the mind incongruous judgments throw the machinery of reason out of gear. So many judgments have been found fallacious by every individual in the race of men, and fallacious judgments have led to such dire disasters, and have been repeated so often in matters of profound moment, as well as in matters of superficial consequence, that there has grown up a habit of mind by which incongruity of judgments is taken as a signal that danger lurks in the way. The mind cannot rest content with an incongruity. It is the ultimate spur to all intellectual activity, for we may forego the pleasures of the mind when we know that others may be enjoyed, but oftentimes we cannot neglect the dangers of false judgments. We must make a practical solution of every incongruous judgment at the time, but every intelligent man yearns for an ultimate solution, thus the world is on the *qui vive* for knowledge as for the breath of life. Those who teach the doctrine of the unknowable offer stones for bread and vipers for fish.

All our concepts must be congruous; the demand for congruity is inexorable. A man may accept a verbal explanation of the facts of science and believe that he has a world of congruous concepts, but experience will find incongruity, which he may conceal for himself in a jugglery of words, but others will detect it when they are announced.

This final faculty in verification resorts to the multitudinous concepts of which the mind is possessed and when one is incongruous with others it demands a reinvestigation of that one. Sometimes the one is right and the many are wrong, and the multitude must be made to establish congruity with one, but meanwhile the one multiplies until it becomes the many and the fallacious judgments the few.

All scientific research is a process of reinvestigating our concepts and of adjusting them to the light which has been shed upon them by some broader generalization than we have been wont to make. We gain a concept by induction and immediately we apply it in a multitude of ways by deduction, and in making these applications we discover our fallacious judgments and go on forever to readjust our concepts. Thus there is trial and failure, trial and failure, until at last there is trial and success; then a new vista is opened into the universe.

The sensations, perceptions, apprehensions, reflections, and ideations of the individual are not exhausted by an enumeration of these derived by the individual in his converse with nature. From his ancestors he inherits the powers of thought, with his organism, which is expectant and apt in judgment and conception. It is ready for this work, as it has been developed through untold generations of ancestral life, and apt, as it has been trained by the experience of untold ages. With the power and skill thus developed it is able to deal with and rationally idealize an immeasurable body of facts which it cannot discover for itself—facts gathered in other lands by other minds and conveyed to it by the agency of language.

The landsman may learn from the mariner, the dweller in the valley from the mountaineer, the denizen of the forest from the denizen of the prairie, and he who dwells where tropical hurricanes wash the coral reefs with the waves of the sea, may learn from him who dwells among the cliffs of ice and sees the bergs of crystal plunged from their glacial homes into the depths of the sea. This process of forming judgments we call acception.

In converse with nature, man transforms or interprets symbols of sense impressions into concepts of sensation, perception, apprehension, reflection, and ideation. In contact with these natural symbols he devises a new world of symbols with which he interprets concepts of others. Still they are judgments founded upon the five factors or constituents of bodies, and nothing more enters into them. So we still find mind dealing with number, space, motion, time, and judgment, or their reciprocals kind, form, force, causation, and concept.

Words themselves are of great assistance to ideation in that they symbolize with one word great groups of judgments which we call concepts. Thus it is that the ego is diverted from the material world to the ideal world, and caused to dwell abstractly upon judgments and their compounds. Perhaps abstraction is more nearly complete in the consideration of judgments and their compounds than in the consideration of times and their compounds, motions and their compounds, spaces and their compounds, and numbers and their compounds. In fact, this abstraction is so thorough that conception is often supposed to have perfect independence of matter, although no conception or judgment is known which is not a concomitant of matter.

A crude speech is developed by all animal life— a general sign language by which every animal holds converse with the members of its own species. This general sign language is inherited by man and gradually developed by him; but oral speech soon leads the way in the development of a still higher language. This oral language is invented by minute increments born of experience; finally, written language is developed from lowly beginnings in picture writings—first, words are developed, and these words are grouped in sentences, and this grouping reacts upon the words themselves until parts of speech are developed, for, in primeval languages, there are no parts of speech as organs of the sentence, as we now understand this term.

Words are signs of concepts, not of judgments, for every word stands for an assemblage of judgments, and to express a judgment it is necessary to formulate a proposition. Yet we cannot get away from sensation, perception, apprehension, reflection, and ideation. The words themselves are spoken or written, and sense impressions are necessary to produce the changes in self upon which consciousness is founded, for consciousness, as we use the term, is awareness of change in self. Thus the spoken word is a sound impression upon the organ of hearing; the written word a light impression upon the organ of vision, and the impression becomes a symbol for sensation, or a symbol for perception, or a symbol for apprehension, or a symbol for reflection, or a symbol for ideation. So all words are symbols for ideation, but the symbols are conventional—invented by mankind for the purpose as an addition to the natural symbols. Not that languages are invented as fully developed, but the elements of every language and the combinations of these elements are invented by minute increments. To understand the word itself it is necessary that there shall be a consciousness and an inference leading to a judgment that the word is such or such, as a sound or a written symbol, and the whole process by sensation and perception must be repeated with every word in order to distinguish it as a word. Then perception, apprehension, and reflection are all employed in confirming a judgment about the meaning of the word, and no word has any meaning until it is interpreted into concepts of number, space, motion, time, or judgment, one or all.

Here we have especially to note that acception becomes not the sole but the chief agency for the development of the concepts of mind.

And now on symbol wings as magical words, the soul flies to all the realms of the universe, learning not only of the worlds of space and time, but penetrating into the arcana of other souls.

By the invention of speech man has acquired an inexhaustible resource from which to draw ideas, but by this artificial method dangers are involved. Imagination often outruns the ideas expressed in words, producing illusions, but usually harmless illusions. My friend tells me of a cove carpeted with rare flowers. I listen and in my mind a brook tumbles in a cascade from a cliff above and the cove seems a deep narrow gorge with fringing rocks and trees standing at the foot of the cliffs. I even perceive in my fancy the pathway by which it is reached, and measure off its distance in my mind's eye. Unexpectedly we come upon the brook. I had imagined it to be much farther off. Thus I had misinterpreted the statement of my friend. We turn up by its bank into the glen. As we enter the cove, instead of finding a narrow glen, with towering walls and overhanging rocks, I see a stretch of pasturage land inclosed by rocks that are broken back in hills, and up the valley beyond the pasturage lands there is a deserted cabin. Near the cabin a great spring gushes from the foot of the rock, and about it trees grow. While my companion gathers flowers I muse. How strange that his words created so vivid a picture in my mind, and that this picture should be wholly the creation of my own imagination, having no counterpart in the reality! I fancied a narrow cove with towering cliffs, tall trees, and a cataract. It is a semicircular glen with broken walls of rock, grass-land and a great spring.

Words are signs of ideas to be interpreted by the imagination of the hearer, and a true or a false interpretation may be given them, depending upon the knowledge already existing in the mind of the hearer.

There is a constant tendency to learn words without meanings, or words with vague meanings, and to use them with a semblance of expressing ideas. No word is properly understood when it does not stand for an idea about one or more of the concomitants of body or about the relations of these concomitants. Here we have a crucial test for the legitimate use of a word; if it does not express a number, a space, a motion, a time, or a judgment, or their reciprocals as kind, form, force, causation, or concept of a body or a relation of one body to another, it expresses a pseudo-idea. A word used to express an idea of an unknown thing may become legitimate by the unknown becoming known, but a word used to express an unknowable thing is blank voice. The habit of learning words without learning the ideas for which they stand is worse than an inanity—it is a vice, for the mind is irresistibly led into the practice of informing such words with vague and misleading meanings. Select any word in common usage to express the leading ideas in the metaphysical discussion of the nature of the universe, and follow it where it occurs many times, and you can invariably discover that it is used with many meanings wholly incompatible with one another, and the foundation of these meanings will be discovered to be something unknowable—a nothing, an abstract attribute reified as having concrete existence. Teach the word cat to a child who has never seen a cat and it will imagine a hobgoblin.

Words often have many meanings; learn these many meanings of many concepts, put them together as one compound idea and you have an absurdity; but such is often the method of metaphysical reasoning. Akin to this is to use the word as a metaphor and then to forget the metaphor. See how Hegel uses the word mediation. A mediator is one who comes between others; the ether mediates the light between the sun and the earth; the air mediates the sound between the voice and the ear; so the messenger mediates the message, and the term properly means to bear from one to another. A man may bear his own letter to his friend and by a figure of speech may be said to be his own mediator. But when you forget the figure of speech and call the man a mediator who acts upon another you have used the word illegitimately, and when you go still further to speak of the action of a person upon himself as mediation, you have reduced the term to an absurdity. Such are the methods of ontologic reasoning as distinguished from scientific reasoning, which holds words to single and invariable meanings. "If thine eye be single the whole body is full of light." It is not strange that Hegel rendered the world into terms of multitudinous contradictions. It was the trick of tricks, the juggle of juggles, to play such pranks with the terms of philosophy.

CHAPTER XIX

INTELLECTIONS

I shall now review the doctrines set forth in these chapters on the five faculties and make a more comprehensive statement of certain fundamental principles.

In this volume psychology is treated only as a system of intellections, while the emotions are neglected. The subject matter is the beginning of an epistomology or theory of cognition, which will require another volume for its completion, when a volume of psychology will follow.

It has been set forth that consciousness is self-consciousness. When the self is conscious of an effect on self it infers a cause, and when it is conscious of being a cause it infers an effect. In the simplest judgment causation is involved—one of the terms being a cause, the other an effect. When consciousness is of the effect, the inference is of the cause, and we have a judgment of intellection. When the consciousness is of the cause, the inference is of the effect, and we have a judgment of emotion. When the cause and the effect are both internal we have an emotion. I use the term consciousness solely as awareness of self and not in its general signification as cognition. We cannot be conscious of an external object, but we are conscious of our judgments of external objects. In the case of the animate body, which has conscious particles acting on one another, it may be conscious of both cause and effect in the body, because the particles of the body are external to one another, and the ganglia, with their connecting fibrous nerves, constitute the organism by which the consciousness of the particles is ultimately transmitted to the cortex. Thus there is a consciousness of the cortex, a consciousness of the subordinate ganglia, and a consciousness of the particles; so that when the self acts on self there are both consciousness and inference.

The cause at one time is considered as a kind, at another time as a form, at another a force, at another a causation, and at another time as a concept, giving rise to five faculties of intellection, as follows: First, cognition of kind, which is the faculty of sensation; second, cognition of form, which is the faculty of perception; third, cognition of force, which is the faculty of apprehension; fourth, cognition of causation, which is the faculty of reflection; and, fifth, cognition of conception, which is the faculty of ideation.

If this doctrine is true then, fundamentally, we cognize by properties which we find to be concomitant in particles and bodies, and thereby reach a cognition of particles or bodies. It will be seen that judgments are fundamentally abstractions, but that comprehension gives them concrete validity. In the first stage they are judgments, in the second stage they are cognitions.

A judgment is a process of elements. First, there is a consciousness of a sense impression. Second, there is a desire to know its cause; that is, what produced it; what can the impression signify? Third, there is a guess or a choice of some external object as its cause, which revives the consciousness of the concept of the object chosen. Fourth, this second consciousness is compared with the first. Fifth, a judgment is made of likeness or unlikeness between the terms compared. The first cause, when it is sense impression, is an act of something in the environment, but when it is a reproduction it is a self-activity. The second cause is always a self-activity.

All judgments are judgments of cause and effect. The consciousness may be of the effect and the inference of the cause, or the consciousness may be of the cause and the inference of the effect, or the consciousness may be of the effect and of the cause when self acts on self, and then the inference is of their relation, one to the other. Again, both cause and effect may be external, when there will be two judgments, each one of which will contain a consciousness and an inference, and their relation to each other as cause and effect will be by inference. Thus inference may be in the second, or higher degree.

There are two of the psychic elements in a judgment that demand further consideration. These are consciousness and choice. Here consciousness is awareness of an effect, the cause of which is an act of the external world thrust upon self at the present time, or upon self at sundry past times. The inference, or interpretation, is a choice of, or guess at, the cause. Thus consciousness is of self, but choice or inference is of the object.

I have spoken of the choice as a guess or an hypothesis, but in cognition it is always an invention, and as an invention it requires the conscious time of deliberation. The mind always invents the cause, and it is because it is an invention that it must be verified; but in recognition the invention is already made and the process of judgment no longer requires deliberation. It is this absence of deliberation which makes multitudes of judgments practically instantaneous, or intuitive. No scientific man can make practical additions to knowledge who is not an inventor of hypotheses. One of the *sine qua non* conditions of successful research is the power of inventing hypotheses; another of these *sine qua non* conditions is verification, but experimental verification also requires invention.

That a primitive judgment requires much time is learned only by careful introspection. So many of our judgments are recognitional instead of being cognitional, that judgments usually appear to be instantaneous. In defense of this doctrine I may be permitted to cite my personal experience. For many years I was engaged on an exploring expedition where all the features of the landscape were new to me and my companions. Mountains, hills, rocks, plains, valleys, streams, all were new. I was constantly discoverng new plants, new animals, and strange human beings, as Indians. During all these years the fundamental doctrines of psychology often constituted the theme of my thoughts and the subject with which I beguiled the weariness of travel. It was thus that I learned to distinguish the elements of a primordial judgment and to distinguish cognitional from recognitional judgments. In later years reconnoissance was developed into survey, and my time was devoted largely to structural geology. For every phenomenon there was always a hypothetical explanation, and such guesses were all found valueless unless verified. Thus it was that the doctrine of a primary judgment and the doctrine of verification grew up with me. More than that, I discovered that my associates in the work of research depended upon hypothesis and verification; and before my field work was done the universal doctrine of cognition herein presented was abundantly confirmed.

Let us look further into the judgment relating to cause and effect in the external world. A judgment about an object may be combined with another judgment about another object, and a third judgment of causation arises, which is about things objective. For example, I see a man strike another and cause pain in that other. I must make two judgments of perception to see the men, a judgment of understanding to see the force, a judgment of reflection to see the effect, and perhaps another judgment of perception to realize the pain. While this maneuver is passing in the field many other events are occurring under the eye, the ear, and other senses, and the many judgments combine in the verification of the judgment formed of the maneuver. Thus judgments are verified. See what a number of concepts are aroused in this case and how much more complex it is than a simpler judgment of sensation, or even of perception.

What we have to note here is the distinction which has to be made between a judgment of causation, which is a highly compound judgment, and the part which causation plays in all judgments—even the simplest.

A judgment of causation is a very distinct thing from the property of causation in the formation of a judgment. In a judgment of sensation I reach a conclusion about a property, say of taste, but I do not consider the cause as a cause but as a kind. So, in a judgment of perception I consider the cause as a form; in a judgment of apprehension I consider the cause as a force; but in the judgment of causation I consider the cause as a cause. Now, the very same phenomenon may be considered in any one of these lights, but the inference in the several cases will be different and the concepts aroused will be different. Which one of the judgments will be made will depend on my interest or the line of thought on which I am engaged. The reader can not be too careful in thoroughly mastering the distinctions between the five classes of judgment, and between the rôle of causation in making a judgment, and a judgment of causation. In the judgment of sensation I think about the kind; in the judgment of perception I think about the form; in the judgment of apprehension I think about the force; but in the judgment of reflection I think about the causation as cause or effect.

Let us now see how cognition is judgment and verification. What things are necessary that I may know that a body has touched me? First, I must be conscious of an effect on self; second, I must infer that something exercises a force that must have produced this change by collision, and the something is a cause and I have a judgment. This judgment may then be verified by my vision when I see the body. The change in self may have been produced by an irritation of the skin due to some disease. What I supposed to have been touch might have been an illusion, but seeing the body as it touched me the verification is made and a certitude is produced. Again, I might have seen the body approach and feared that it would touch me, and expectant of the touch, I might have inferred the touch when really the touch was not accomplished. In this case there was a consciousness by the sense impression in vision, but an inference which was only an illusion. Two or more acts of consciousness producing the same judgment verify one another.

How must I know that a knife has cut me? First, I am conscious of a change or effect in self; second, I infer that something has produced that effect as a force and I have a judgment. In order that the cognition may be complete this hypothesis must be verified. I may verify the cutting by seeing that the gash is made, and I may verify the knife by seeing or touching the knife. In the one case I have a certitude that I have been cut, and in the second case, a certitude that I have been cut with the knife, and these certitudes verify each other. I might have seen the knife move near to my hand and inferred that it cut me, and an illusion might thus have arisen that I was cut; but the consciousness of the effect of the knife upon my hand, together with the consciousness of the knife by vision, produce judgments that confirm each other.

How do I recognize that some one has spoken? First, I am conscious of a change in my organ of hearing. I infer that it is the sound of a voice. I see the person's lips move, and it is confirmed and I have a certitude. I might have seen the lips move without hearing the sound, and inferred that a sound was made, which would have been an illusion; but in the hearing of the sound and the seeing of the movement of the lips, each verifies the other.

How do I cognize the flavor of an apple? My taste and my vision of the apple verify each other. How do I cognize the odor of a rose? By smelling and seeing, and the common judgment is verified.

In these cases judgments verify one another. All verification is founded on congruence of judgments. It is thus that one sense verifies another. Now, that which we have specially to note at this stage of the argument is that verification is founded on congruence of judgments. Every cognition involves a judgment and its verification, and the verification is founded on the congruence of judgments, one with another of a higher grade.

In the lower stages of the development of mind, verification is sometimes by repetition, oftener by submitting the judgment to verification by another sense; but in the higher stages of the development of mind, verification is by experimentation. We go on from generation to generation with unverified judgments and suppose that our concepts are composed of cognitions, when in fact they are composed of fallacious judgments. For untold generations men believed the earth to be flat, and that bodies fall to the earth in a line normal to this flat plain. But there were certain phenomena which were inexplicable, and men invented the hypothesis that the earth is a spheroid and that bodies fall toward the center of the earth, and it accounted for so many facts relating to the motions of the heavenly bodies that the hypothesis led to a vast amount of scientific research, and was verified. Now at last we cognize the motion of the heavenly bodies in part at least. For ages man believed the heavenly bodies to be molar, that is, to be movable by man at his will, moving from east to west along the face of a solid domed sky, and they supposed them to return from west to east under the ground, and it required ages to invent another hypothesis—that of a system of spheres revolving in orbits concentric to the sun. This hypothesis was an invention made by many men. It was a demotic, not an individual invention.

The various judgments formed about an external object are combined into a concept of that object, and this concept is aroused from memory by inference whenever a sense impression is received and attention is paid to it in judgment. One sense impression becomes an agency for reviving many judgments previously made about the object causing the sense impression. It is thus that a sense impression becomes symbolic, and judgment in such cases is symbolic. The concomitant properties of an object severally manifest themselves to different senses, and when one property is manifested by one sense impression, it becomes the symbol of all other properties inhering in the object and known by the observer. Properties can not exist apart, as the constant multitudinous experiences of each individual attest. There is no one who can form a judgment who does not take it for granted that the concomitants, however unlike they may be, can not exist apart. Symbolism is not mere poetry that obscures reason, but it is a logical method of time-saving thought. Judgment itself is by symbolism, in which the manifestation of one property is interpreted as a symbol of all the properties known about the object.

A force is manifested as a force and it is also manifested as a cause, for there can not be a force without it also being a cause, any more than there can be a force which is not a form, or a form which is not a kind. In nature forces are often observed in multitudes. There are many particles of air that stir the leaf and there are many leaves that are stirred by one wind, but in the particles of the wind one multitude follows another in succession. So there are many drops of rain that fall on many grains of soil, and a succession of a multitude of raindrops constitute the rain. Process in its simplest form is the collision of two bodies that meet and act on each other in action and reaction, but this action and reaction is also cause and effect; thus causation and force are concomitant. But in apprehension we consider only force; another intellectual faculty is engaged when we consider causation.

When one body collides with another, different things may happen. First, both may be deflected; second, both may be deformed; third, both may be broken; fourth, both may be heated; fifth, both may chemically be changed. Usually the total effect is two or more of these changes. Finally, any one of these effects may be experienced by one body and not by the other. Thus we see that although action and reaction are equal, cause and effect can not be equal, as they are not of the same kind.

Judgments of reflection seem to be especially subject to error and as such to be compounded into false concepts and to be long entertained as such. In the act of making the judgment there must be judgments of bodies impinging on one another, leading to judgments of apprehension. Then one of many effects must be considered as due to one of many causes, and the many effects referred to the many causes in turn, in order that all of the effects may properly be distributed to all of the causes. Thus reflection is an exceedingly complex subject.

The process is comparatively simple when one body collides with another, but when a multitude of bodies collide with one, the process is not so readily understood, and when a multitude of bodies collide against a multitude of bodies, as of winds against leaves, the process of disentangling causes and effects or antecedents and consequents, is still more involved. The difficulty may not appear at first glance, but an investigation into historical instances shows that frequently cause is mistaken for effect and effect for cause. It is not uncommon in savagery to attribute winds to trees. A common error of this kind is discovered in the minds of most persons, for it is widely believed that forests are the cause of rains. An interesting book has been written, widely read, and popularly approved, which is based on the assumption that the aridity of desert lands is due to the absence of forests.

A stream of judgments flow through the mind. As the ego has self-activity it changes its position in the environment at will and a different environment plays on the senses at every change in the position of the ego. Then by different senses the environment solicits the attention simultaneously by all. Thus attention is solicited by more sense impressions than it can attend to, and it chooses for attention those which serve a temporary or more sustained purpose. Those serving a temporary purpose give rise to what has been called by Kant, the practical reason; those serving a sustained purpose, the pure reason.

Presentative judgments that originate in sense impressions, are often followed by representative judgments, and these are either discursive or volitional. Hence we see that the judgments which we make are exceedingly multitudinous and heterogeneous. But all of these judgments are assembled in concepts by more temporary or more permanent purposes. What judgments can be made are determined by the environment; but what judgments the mind selects to make are determined by the purpose. Thus the ego is the creature of environment and self-activity. The stream of judgments is thought, and thought is controlled by self-activity and environment.

It may be well to further consider the process of combining judgments by reflection.

I am wandering by the river. Why should the river here suddenly pass from a narrow gorge to a wide-spread plain and be transformed from a narrow to an expansive stream? And why should the turbulent waters above become so quiet below?

I climb a rock to study the problem. The bluffs standing back from the river, converge at this point and seem as if they would join hands across the chasm through which the river plunges. Here the bluff is a cliff and the edges of sandstone strata outcrop in the escarpment, and I observe with care the succession of rocks from the bottom to the top of the cliff. But a robin flies down and perches on a willow near by, and in an instant cliff and geology vanish from my thought; I see a turkis egg and a nest in the apple-tree of my garden, and my daughter is shouting a song of childish joy in my mind's ear, for this she did, not many weeks ago.

In thought I am at home once more. Then home vanishes and I see the robin again flitting from bough to bough, and as it moves my eyes follow it until it is in a line between myself and the cliff, and the sight of the cliff brings back my geologic problem. I see the red sandstone below, the brown shales between and the white sandstones above, and recognize the succession as being similar to one seen before. If so, the summit of the cliff must be crowned by a limestone. Yes, there is the limestone with its angular outlines, in contrast with the round reliefs of the sandstone. I am one step farther in my problem. I put the facts of the succession together and say this is a Carboniferous cliff. I know these rocks.

In climbing I hear a noise. In an instant I interpret it as the voice of a friend, and turning about, find I am right. I hasten to announce my discovery, but he holds a flower aloft, waving it in triumph. That wand banishes the cliff with its succession of beds from my mind, and I see a bluebell drooping from its delicate stem and ringing a chime of cerulean beauty. In a twinkling of an eye my mind travels a thousand miles, and I am climbing the gray sandstone cliff which rises in the midst of the valley of Illinois river and is known as "Starved Rock." The miles my soul has traveled are only equaled by the time over which it has returned. I am a young man again, and I burst into a song:

> "It's rare to see the morning bleeze
> Like a bonfire frae the sea."

Why do I sing that song? It was on my tongue when I found my first bluebell on "Starved Rock."

My friend bids me follow him. At one moment I am thinking of the cove, at another I am listening to the voice of my friend, and at still another I am watching the way over which we walk; and now and then my mind wanders away home and where not. Now my attention is attracted to a footprint in the sand. From its shape I know it was made by a deer. Thus I make an inference beyond my perception. The track is the sign of something else. I see other tracks; they are arranged along our course in pairs several feet apart. By this arrangement I infer that the deer was leaping, as if fleeing from danger, and I imagine that the deer has been startled at our approach. This is an erroneous inference, for my friend tells me that he roused the deer as he came down the path some time ago. And as we still walk I study the rocks, and discover that a limestone forms the floor of the valley below; and then I discover by its contained fossils that it is the same formation as the one which crosses the summit of the cliff. The valley limestone was broken from the cliff limestone and dropped down by what geologists call a fault, and the fall or throw of the fault is more than a thousand feet. And now I discover the origin of the cascades in the canyon above and the broad and quiet flow of the river below. The last dropping of the sandstone by the fault decreased the declivity of the stream in the valley and increased the declivity of the stream above the valley, where it comes down through the canyon. All this is reasoning. It is a series of judgments controlled by will for a course of reasoning on a theme for which I have a permanent interest, interrupted by a multitude of adventitious judgments that are made by reason of temporary interest.

We sit down by the spring and my friend spreads the lunch on a fallen tree trunk, and away goes my mind to the bank of the Grand river in central Colorado, and I see a prostrate pine, and an emerald lake near by, and on the shore, cliffs of granite, and beyond, a snow-clad mountain, and about its summit the gathered clouds, and the sheen of clouds and snow-fields blends with stretches of forest and crags and peaks of towering grandeur. Years ago I was there, and the feast on this log brings back the feast on that log, with its attendant glories of mountain scenery. From that scene I am called back by the bidding of my friend to eat. Then a bird comes down to the fountain, and I am engaged in watching its coy advances to the water. And so my mind passes instantaneously from one object to another—now engaged in observing things present, now listening to the voice of my friend, now occupied in expressing my thought to him, now calling up some scene from afar; but ever thinking. On goes the stream of thought.

I eat of the turnover, and observe from the taste that it is made of blackberries; and then I think of the blackberry patches over which I strayed in childhood on the hills of southern Ohio, and of my companion, Charles Isham, who was killed at the battle of Shiloh. And I talk of battles, till my friend speaks of bread and butter. Thirst causes me to go to the spring, and I quaff from its crystal fountain, and listen to the jests hurled at me by my friend, and laugh at his wit. Still on goes the stream of thought.

We have eaten the lunch and gathered the plants, and return home. On the way a sharp, buzzing sound thrills me with horror. I know it as the warning of a rattlesnake. It is a familiar sound to me, for I have found many of these serpents in the wilderness. I look about, and there it is, coiled in the grass. With my cane I strike it a blow, and then another, until it stretches its length on the ground, dead. From the inanimate reptile I pluck the rattles. There are nine on its tail, which it was wont to ring when danger approached—discordant bells whose ringing is a symbol to the woodsman that reptilian hell is lurking near the pathway.

We have reached the river bank, and separate; I climb about it in search of fossils. Soon I discover carboniferous fossils in the rock at the foot of the cliff, and climbing up beside the stream I discover limestone rocks which have come down from the summit of the cliff, and see the same fossils. My explanation of the origin of the cliff, the rapid descent of the river from above, the narrow channel through which it runs, the valley below, and the broad expanse of quiet water, is verified. Now, in my reasoning about the fall of a river into a quiet reach, I used concepts of form in the nature of the channel, and concepts of form in the structure of the rocks. I also used concepts of time in the succession of the rocks, and I reached a conclusion or judgment as to the cause of the rapid which was a judgment of causation, and I confirmed this judgment by reaching the same conclusion from the story of the fossils that I had reached from the story of the geological structure; so concepts verify concepts. On careful examination it will always be found that judgments of causation are verified by the congruence of concepts.

The stream of thought is composed of a series of widely diverse elements, or mentations, that are judgments, all differing among themselves. Now, it is impossible for the mind to dwell on any one of these elements. You cannot think of a scratch long; the mind immediately passes to something else—another sight or sound. Consciousness, which is awareness of a change in self, is the absolute, the independent of thought and that on which inferences are founded; and consciousness is awareness of a succession of impulses on self or by self, that flow with the rapidity of thought that seems almost to vie with the rapidity of air collisions in sound. Hence consciousness is serial, and inferences are serial, and judgments are necessarily serial; but thought must go on. Gaze into the eye of my lady and think of its sapphirine hue; in a moment you think of something else—the sable curtain, the coy glance, perchance the cerulean heaven, or the deep blue sea. It is impossible to hold your mind for more than a moment on the blueness of the eye; the thought must go on. But on to what? is the question. Tell me in the case of any individual the laws which govern the procession of his thought, and I will tell his name, be it sage or fool. There is always a nexus between contiguous elements in the stream of thought. Sometimes it is mere adventitious association. The thing seen or heard has at some previous time been associated with something else. The touch is associated with the mother's stroke on childish curls; the taste of that particular fruit is associated with an occasion of joy; the perfume of smoke is associated with the burning forest; the song is associated with some scene of glee; the robin is associated with the cottage home. But the nexus of association is not always adventitious. It is often controlled by an established design. With the fool, adventitious relation is the principal nexus of thought in the procession; with the sage, logical relation is the chief nexus.

The links of relation in the chain of thought are not always apparent to the thinker himself. Steps in the procession of reasoning are often canceled; the mind passes, by great bounds, from one to another. When the steps in the course of logical reasoning have been taken many times, the mind finds it unnecessary to tread the ground again and again, with slow and measured pace, but it springs from point to point, and the greater reasoners make the greater leaps. This is a fact well known to scientific men, but it gives to the procession of mentations those characteristics which cause the greatest wonder to men, and which have led to many of the errors of psychology.

By reflecting on the past and comparing it with the present, we prophesy of the future and often our prophecies are confirmed. By day we prophesy of the night, and the night comes; at night we prophesy of the morning, and the morning comes. As the days, weeks, months, and years go by we learn by experience of the changes wrought in self and infer changes yet to be wrought. By experience we discover the changes wrought in others, and by inference judgments are formed of changes yet to be wrought. It is by experience that we learn of all the changes in environment. The skies change; the seasons change; the river was low yesterday, it is a raging torrent today. The acorn bourgeons with leaflets, it sends rootlets into the earth and stem and branch into the air; it grows from week to week, month to month, year to year, and under our experience it becomes a tree. The child is born, it grows to be a lad, a youth, a young man, a vigorous adult, an old man, and the judgments formed are compounded into ideas of becoming. It is thus by reflection that a vast multitude of judgments are compounded into ideas of the changes wrought by time, and reflection becomes the special process of cognizing metagenesis. As on the wings of perception all lands are viewed, so on wings of reflection all times are conned. The illimitable past and the illimitable future are all painted on the canvas of now by the artist of reflection. Things that have been and things to be are emblazoned on the panorama of reflectional concept.

Thus we have ideas of sensation or classification, ideas of perception or integration, ideas of understanding or coöperation, and ideas of reflection or history, all derived from the germs of sense impression as they have been made on the mind of the individual in moments, hours, days, and years.

A boulder cannot move from the bank into the swift channel in order that it may journey down the stream, but a man may travel from the distant hill to voyage on the river. The leaf cannot flutter in the air unless the air is sweeping by, and the air cannot move as a breeze without antecedent conditions of temperature. Every action is self-action and every passion is self-passion, but the action of one must have its correlate in the action of another, and the passion of one must have its correlate in the passion, of another. In this respect animate bodies have a property which separates them from inanimate bodies, in that they perform actions which are self-directed, and in that they have passions that are self-chosen. The animal may choose to enter the current or it may choose to expose itself to the wind, and it may act for these purposes by placing itself under the proper conditions. Heretofore we have attempted to use the term activity in this sense as a chosen act. By such activities design or purpose is expressed. I see a bird fly from tree to tree and think of it as an activity prompted by design. I see a leaf blown from one tree to another and I see an act not determined by choice. All this is intended to make clear the distinction between activities and acts and to show that activities are manifestations of mind. The animate body is conscious of mind, and through the manifestations of mind with others it is led to infer that they also have minds.

In the history of metaphysical philosophy the doctrine of presentative and representative judgments has undergone some strange vicissitudes. The distinction seems first to have been formulated by the terms impressions and thoughts, presentative judgments being called impressions and representative judgments thoughts. Spencer refers to the same distinction when he speaks of vivid impressions and faint impressions. Others have considered presentative judgments as instinctive or intuitive, for such judgments are often made instantaneously and without apparent consciousness of previous judgments. The nature of intuition we have already set forth. Kant also believes that representative judgments are controlled by forms of thought preëxisting in the mind and not derived from experience, in which all judgments are molded. He supposes the mind to be endowed with the knowledge of space as empty space and of time as empty time, and that the ego fills the empty space and empty time with forms of thought. Thus the metaphysicians have always failed to discover the nature of a judgment with its pentalogic elements, in which both consciousness and choice appear with comparison, which completes the judgment. They also fail to discover that a presentative judgment is only initiated by a sense impression, and that the ego must still recall past impressions in a concept to make the judgment complete, and they also fail to discover that the representative judgment is initiated by recalling a past concept and comparing it with another concept of past judgments.

I see a worm crawling on the ground; the worm causes a sense impression. I might stop to consider its color and have a judgment of sensation, or I might consider its form and have a judgment of perception, or I might consider its motion and have a judgment of understanding, or I might consider its cause as an egg and have a judgment of reflection, or I might consider that the motion itself is directed molar motion and hence manifests mind in the worm; then I would have a judgment of ideation. Any one of these judgments can be made from the same sense impression, and my interest, my purpose, my choice determines the nature of the judgment made. But when made it needs verification.

If the judgment as a sensation is valid and there is a color, if the judgment of perception is valid and there is a form, if the judgment of understanding is valid and there is a motion, if the judgment of causation is valid and there is an object developed from an egg, then there is left for consideration the validity of the judgment of ideation, for the worm may not be moving by its own volition but it may be dragged by an ant. Its motion must be due to an animate and designing cause, which may inhere in the worm itself or in another which is unknown to me, for it is molar motion caused by mind, and in order that I may verify my judgment of mind in the worm I must determine that it is living and free to use its own judgment; such verification comes only by the comparison of concepts. As ideation is the compounding of concepts, so verification in ideation is the comparison of concepts.

In sensation, perception, understanding, and reflection, concepts are developed by the consolidation of judgments. In ideation we have a faculty by which judgments are added to judgments to constitute concepts and which then continues its power of forming judgments by combining concepts with concepts and forever forming new concepts thereby, while at the same time the power thus developed of comparing concepts with concepts is leading to a re-formation of the concepts themselves by the elimination of fallacies, for when concepts by comparison with concepts are found to be incongruous, the mind refuses to accept them as valid and seeks for the source of error. We must, therefore, discover the means by which concepts are compared with concepts.

We must now shoulder the task of explaining the laws of symbolism or association, which have been assumed from time to time and partially explained in this discussion.

It has been shown how concepts are formed as groups of judgments in sensation, perception, apprehension, and reflection, and how ideas develop simultaneously. We are now to show how they are compounded with one another, and how in this process incongruous ideas are adjusted by the elimination of judgments that are fallacies, for judgments must ultimately die if they do not fit in their proper places.

That which I have sometimes called symbolism and that which I have sometimes called association are the same thing. When a sensation which is the result of a sense impression caused by one attribute of a body, is taken as a symbol of the body itself with all its attributes, it becomes a symbol of all with which it is associated. When a sense impression gives rise to a judgment of force it recalls many other judgments of force and thus becomes a symbol of other things. When a judgment of cause is formed it also becomes a symbol of other causes. Sense impressions are directly used by the mind in this manner in sensation, perception, apprehension, reflection, and ideation, and it is thus that ideas are primarily associated. The memories of judgments are recalled by other judgments, as we have seen, so that not only do judgments which arise from sensations recall other judgments, but these other judgments recall still other judgments, and thus there is recollection in the second degree; and these revivals may go on from degree to degree to an indefinite extent. All of these facts have been illustrated.

As we judge by comparing concepts with other concepts or with impressions, one judgment by a faculty is associated with other judgments by the same faculty, and as one property is concomitant with all the others, one property becomes a symbol of all the others, so that there is association by comparison of concepts and association by symbolism. Hence all our judgments are associated.

The quantitative properties are the reciprocals of the categoric properties, for the one is the reciprocal of the many which compose the one. The one is a kind, and the many is another kind, and the one kind is the reciprocal of the many kinds. So the one form of the body is the reciprocal of the many extensions of the particles. The one motion of the body is the reciprocal of the many motions of the particles, hence the one force of the body is the reciprocal of the many motions of the particles, for the force of the body is the reciprocal of the motion of the particles. The one time of the body is the reciprocal of the many times of a particle, hence the one causation of the body is the reciprocal of the many times of the particles. The one judgment of the body is the reciprocal of the many judgments of the particles, hence the one concept of the body is the reciprocal of the many judgments of the particles.

Judgments of quantitative bodies are reciprocal judgments of classific bodies, hence they are associated by reciprocality. Judgments of one property are concomitant with judgments of another property, therefore they are associated by concomitancy. Now judgments associated by concomitancy are often intuitive in the sense in which that term is used here; so judgments associated by reciprocality are often intuitive. But there are many judgments that are associated not by concomitancy or reciprocality, because they are chosen when we make judgments; of those chosen some are volitional, some discursive. The discursive associations are those usually recognized as such, and again we have association by kind or likeness, by form, by force, by causation, and by concept. Thus it is that the ego remembers by pentalogic properties. Thus association is the law of memory.

Units are associated with units, numbers with numbers, kinds with kinds, series with series, classes with classes, and all are associated in nature and considered in classification. Then extensions are associated with extensions, spaces with spaces, forms with forms, metamorphoses with metamorphoses, organisms with organisms, and all these are interassociated and these associations are considered in morphology. Then speeds are associated with speeds, motions with motions, forces with forces, energies with energies, powers with powers, coöperations with coöperations, and all of these modes of motion are interrelated or associated and all are considered in dynamics. Again persistencies are associated with persistencies, times with times, causations with causations, metageneses with metageneses, developments with developments, and they are all interrelated and considered in evolution. Finally, sensations are associated with sensations, perceptions with perceptions, apprehensions with apprehensions, reflections with reflections, and ideations with ideations, and all are considered in intellection and are represented by words. Then numbers, spaces, motions, times, and judgments are associated, and kinds, forms, forces, causations, and concepts are associated, and the quantitative properties are associated with the categoric properties. There is a congeries of associations in which all of the contents of the mind are associated as fast as we cognize the bodies of the universe in their properties and relations.

Certain special associations of discursive thought have received special attention and various attempts have been made to account for them, while the multitudinous associations of thought have been neglected. This partial discussion of the subject has led to the classification of the associations of memory and two laws have been formulated: the one called the law of likeness, and the other the law of contiguity. They have also been formulated as three or more; but the essential nature of association has failed to receive attention because the five associated properties of matter have not clearly been understood; all of these methods, about which scarcely two psychologists agree, have been inadequate to properly set forth the subject. Especially do we notice that contiguity in space has been confounded with immediate succession in time by the habit of using a word with two meanings, and thus, confounding succession with position. Particularly intensive associations by which striking events are recalled, because of the deep effects made on the mind, have been observed by thoughtful men for more than twenty centuries. In moods of contemplation a judgment recalls some remote judgment which was startling at the time, and as we go on from moment to moment, recalling a multitude of things by a multitude of associations, this special instance is thrust on the mind and we stop to consider it. I see a rock which more or less resembles another which I once saw and now recall, together with an event which at that time made an impression on my mind; a man fell over the cliff. I smell the odor of burning brush in the wayside field and I suddenly recall the odor of the fire which I kindled for burning brush-piles on my father's farm. I taste the flavor of a nut and I recall the time when I threw to my shouting companions the walnuts from a wayside tree. Such startling revivals, often repeated, challenge attention, and though thoughtful men have given much attention to the phenomena, it has resulted in a very imperfect psychology of association and symbolism.

Once more the attention of the reader is called to the relations which exist between the five essentials and which are then found in the five properties, then found in the five categories, then found in the five properties of change, then in the five properties of life, then in the five properties of mind. Kinds are not alone classified, but forms, forces, qualities, and concepts are classified. Morphology considers not only forms, but it also considers kinds, forces, causes, and concepts. Dynamics considers not only forces, but it also considers kind, forms, causes, and concepts. Evolution considers not only causes, but it considers kinds, forms, forces, and concepts; and ideation considers not only concepts, but it also considers kinds, forms, forces, and causes, and the difference between these five concomitants is the point of view when every one of the essentials and its derivatives is considered abstractly. As they cannot exist abstractly, the mind cannot overtly consider an abstraction without tacitly informing it with concrete existence.

The error of metaphysic is the confounding of abstraction with analysis by assuming that abstractions have separate existence. If the argument has not made this point clear it has failed of its purpose. The habits of thought engendered by the study of abstract mathematics often leads the mathematician into the very same pitfalls into which the metaphysician stumbles.

The manifestations of properties are symbols, because one becomes the representative of all the others in the body manifested. When animate beings develop the faculty of reading these symbols, they are said to be able to read the expression of the emotions and are themselves expert in the expression of emotions. Gradually these expressions become more and more artificial as animals advance in culture, until at last a conventional language is devised. This is speech, which is practiced by the lower animals, but which is much more highly developed in man. Natural symbolism thus becomes conventional symbolism, and words are signs of concepts. A wholly conventional symbolism is thus devised, the symbols being symbols of concepts. Now, men practically and overtly consider their concepts and a language is a vast reservoir of conventional symbols used for this purpose. There is no human language so crude that it does not have tens of thousands of such symbols, which, put together in propositions or sentences, have the power of expressing all the judgments which the people who use the language are able to make. We now see the enormous development of ideation which man has accomplished by the invention of language.

A judgment is expressed in a proposition by conventional language. Unfortunately, in grammar, subject and object have a different meaning from that which they have in psychology. In grammar the subject means that something about which an affirmation is made, and the predicate means that which is affirmed of the subject, while object has various meanings in grammar. Until the terms of grammar are made to conform with the terms of psychology, there must always be some confusion. Formal logic is the logic of grammar, and the purpose for which it was devised was success in disputation. Scientific logic is the logic of kinds, and it is of scientific logic that we speak in this essay. The logic of which we speak is the logic of reasoning, not the logic of grammar.

The methods of comparing judgments and concepts are innumerable, and every judgment is an act of comparison, and we are forever judging for the purposes of discovering congruities; an incongruous judgment acts upon a healthy mind as a moral irritant. If this and this judgment do not agree, it is an evidence of ignorance and a suggestion of imbecility. There is no other motive that clings to man so long as the desire for wisdom.

CHAPTER XX

FALLACIES OF SENSATION

The certitudes which we have tried to demonstrate have given rise to a host of fallacies which have played a strange rôle in the history of opinions and which from time to time have vitiated science itself. Civilization began with science when it commenced with verification by experimentation. Verification soon led to the dissipation of fallacies; then it was discovered that things are something more than what they seem to be to our simplest judgments. Kinds are something more than kinds, they are forms; forms are something more than forms, they are forces; forces are something more than forces, they are causations; in animate bodies causations are something more than causations, they are concepts. When we know all about a body we must know all of its properties and these can only be discovered by investigation, and science is the result of this investigation; but before we acquire knowledge we entertain fallacies. The early philosophers, discovering that partial knowledge is inadequate to the expression of the whole truth, thought to characterize the whole truth by calling it noumenon, and they thought to characterize partial truth by calling it phenomenon. This was a wise and legitimate distinction, but the time came when certain delusions were held to be sacred and a belief in them necessary to a good life; so they thought by the legerdemain of language to prove that delusions were the noumena and all knowledge only phenomena. But scientific men took up the phenomena or unexplained properties of bodies and by investigation increased knowledge as science, and reduced phenomena or partially explained properties to noumena or more fully explained properties. To a great extent they dropped the term noumenon and held to the term phenomenon, and expressed the opinion tacitly or overtly that a phenomenon is but still a phenomenon whether it be properly or improperly explained, and they held it their province to explain phenomena and they called the explanation of phenomena, science.

In modern times those who hold that noumena are inexplicable, that is, unknown and unknowable properties, call themselves metaphysicians. Those who hold that phenomena are knowable and seek by investigation to know them, call themselves scientists. Such schematizing of philosophers as metaphysicians and scientists is necessarily imperfect, for some philosophers are both metaphysicians and scientists. There are many who are metaphysicians when they wear their holiday dress, and scientists when they wear the garb of labor. Metaphysical reasoning can be more clearly demarcated from scientific reasoning, for scientific reasoning may always be known by its demand for verification. We may make a mistake in sensation because of its obscurity or by referring it to a wrong sense. The sense impression may be obscure itself, as when a sound barely passes the threshold of consciousness, or a sight which is obscure by reason of the twilight, or it may be obscure by reason of preoccupied attention; thus I may fail to attend to a sound or a sight because my attention is elsewhere engaged. I do not hear the speaker because I am attending to a sight, or I do not see a sight because I am listening to what another person is saying. All of such missensations are easily corrected by ordinary methods of verification, but often we neglect them, as we deem them of no importance. I shall call all such errors of judgment, missensations, and group them in a higher class which I shall call illusions.

When a youth, as I was breaking prairie with an ox team, my labor was interrupted by a rattlesnake, and during the day I saw and killed several of these serpents. At one time the lash of my whip flew off. In trying to pick it up I grasped a stick. The fear of being bitten by a snake and the degree of expectant attention to which I was wrought, caused me to interpret the sense impression of touch as caused by a rattlesnake. This was a missensation of touch. At the same time I distinctly heard the rattle of the snake; this was a missensation of audition.

I make a distinction between a sense impression and a feeling impression. A sense impression is one made upon the end organ of a sense by an object exterior to the body; a feeling impression is one made upon an organ of feeling which is metabolic, circulatory, motor, reproductive, or cognitional. A feeling impression arises as a result of the functioning of the organ and is usually distinguished as being subjective. The mind may err in considering a subjective impression as objective, when an hallucination will be produced. We thus divide fallacies of sensation into two groups, missensations and hallucinations. Missensations are easily corrected; hallucinations cannot be corrected while the person who makes them is in the condition of mind under which they originate, for they are produced under abnormal conditions and so long as these conditions prevail similar hallucinations will occur, for hallucinations occur in the dream state, the intoxication state, the disease state, or other abnormal states. We will see the significance of this statement when we proceed to discuss hallucinations. Missensations are at first presentative and they remain only until corrected by verification; hallucinations are false presentations and cannot be tested by the verification of the persons who make them. To the mind that forms the habit of believing in hallucinations they come to the persons as recognitions and have the instantaneous effect of recognitions.

Here we must distinguish clearly between a fallacy of sensation and a fallacy of feeling. A soldier in the suspense which precedes the battle, when sharpshooters are now and then picking off a man, may have his gun or his clothing touched by a rifle ball and in the suspense of the occasion may imagine that he has received a severe, perhaps a deadly wound, and may shriek with pain. The fallacy of being struck is a fallacy of sensation, but the fallacy of having pain is a fallacy of feeling. Similar cases are often witnessed on the frontier, where men experience an adventurous life. Now, we are not treating of fallacies of feeling, but of those of sensation. An hallucination is the antithesis of the one I have just given; it is the error which arises by interpreting a feeling impression as if it were a sense impression; but a fallacy of feeling consists of interpreting a sense impression as a feeling impression.

In a former chapter it was explained that a judgment of intellection is a judgment of the cause of a sense impression, and that a judgment of emotion is a judgment of the effect of an impression. The feelings, therefore, tell of effects upon self, and the senses tell of the causes of these effects. This distinction is important to a clear understanding of the nature of fallacies.

Parish has assembled a great body of "Hallucinations and Illusions," which are in convenient form for reference. As his treatment of the subject is better than any I have elsewhere seen, I shall liberally avail myself of the material which he has gathered. Notwithstanding Parish's disclaimer, he still exhibits a tendency to explain psychological phenomena by a reference to its physiological concomitant. As there can be no psychology without its concomitant physiology, this is quite legitimate, but the practical conclusions at which he arrives still require explication in terms of abstract mind. He uses a geometrical scheme for the purpose of setting forth the facts of physiology. Such a scheme may have an expositional value to make us realize the facts which have been discovered in the anatomy of the nervous system, but it is easily abused. We know that the nervous system is composed of ganglia of cells, connected by nerves composed of bundles of fibers, and that the ganglia are found in hierarchies connected by these nerve fibers, which finally terminate in the organs of life, where they are distributed throughout the system, and also at the periphery, where they terminate in end organs supplied with various mechanical devices. The nerve fibers that connect with a ganglion are not structurally continuous with the cells of the ganglion, so that a sense impression or a feeling impression is conveyed from one ganglion to another by fibers which are discontinuous at the ganglion. This permits of a shunting or diversion of an impulse in many directions through the nervous system, a ganglion being a shunting or diverting mechanism. The paths of which Parish, together with many other authors, speaks, are the fibers and cells. Now, I submit that a simple statement of the fact is much more readily comprehensible than any geometric scheme which any physiologist has devised. The concept of a nervous system composed of sensory and vital organs connected by nervous fibers with nervous cells for a shunting apparatus, is one easily realized by the mind. It must be remembered that this discovery was not available until of late. When we come to explain the physiology of the nervous system we must explain also the anatomy of the nervous system, and finally this leads us to an explanation of the metabolism of the nervous system. Hence conception has its concomitants in physiology, anatomy, and metabolism, and as the physiology of the nerves is a process which also involves time in its evolution, we may characterize conception in terms of evolution, physiology, anatomy, or metabolism, but a psychologic treatment of the subject requires that the conception should ultimately be treated in terms of psychology. I shall, therefore, treat all fallacies in terms of psychology. I shall assume that both sense impressions and feeling impressions may go astray in passing from the end organ to the cortex, because the fibrous nerves are not structurally connected with the ganglionic nerves, so that, under certain conditions, they may be directed to any portion of the cortex by the will acting normally or abnormally.

Every cell in the human body is a seat of consciousness, while the nervous system is the organ of inference. All the bodily organs are related to one another through the structure of the nervous system, the fibers of which permeate all the organs, collect sense and feeling impressions from them, and transmit them by fibrous nerves to the ganglionic nerves, where such impressions are woven into concepts to be ultimately returned to the motor apparatus. In this conception I suppose that an hallucination involves not only the central organ in the cortex, but it also may involve a subordinate ganglion or an organ of sense or feeling.

We have divided fallacies of sensation into missensations and hallucinations. The exposition already made relating to missensations will, perhaps, be sufficient for practical purposes, but hallucinations will require further consideration.

In discussing hallucinations there are no sense impressions to be considered, but there are feeling impressions which are interpreted as if they were sense impressions. The interpretation seems always to be made by the faculty of perception. We have, therefore, to discuss hallucinations as false perception based on feeling impressions; consequently, in order to consider their cause in feeling impressions, we shall illustrate by instances of fallacious perceptions which are specters.

Esquirol distinguishes hallucinations from illusions by considering hallucinations as "subjective sensory images" which arise without the aid of external stimuli, and illusions as the false interpretation of external objects, but he does not clearly distinguish between sensation and perception, which we have attempted to make clear. In the same manner Parish has fallen into confusion; Sully makes the distinction but he classifies illusions in a manner which we cannot follow. I shall therefore treat the subject as demanded by the standpoint obtained in considering the five-fold faculties of the intellect as hitherto set forth.

In sensation we hear sounds that are caused by objective bodies; thus a bell agitates the air and we hear it, but we may have a disturbance of the physiological function of the ear, due it may be to the influence of a drug or perhaps to a disease of the organ. Now, such a subjective impression or functioning of an organ of sense we call a feeling impression, and when we consider it to be objective we hallucinate or have an hallucination.

In a highly nervous state men mistake the motor feeling of speech for the sound of speech, as if caused by another or objective person. A subjective irritation of the skin may be mistaken for the objective crawling of an insect over the skin. A polypus in the nose may produce a disturbance in the function of the nose which is interpreted as an odor. A man may smell paradisic odors or mephitic stenches by reason of disease in the olfactory organ. In the same manner diseases produce hallucinations of the gustatory sense.

The literature of hallucination in large part is the literature of pathology, although the occurrence of hallucinations has often been recorded in biographic literature, in which there are many notable examples.

Socrates had hallucinations of a demon who frequently warned him of impending evil. Savonarola saw the heavens open and a sword appear on which was the inscription *Gladius Domini super terram*. Luther had an auditory hallucination when on the stairs at Rome he heard the words, "The just shall live by faith." Cromwell had his greatness foretold him by an apparition. At first it may be difficult to state whether such fallacies are hallucinations proper or only missensations. As we go on with the subject, however, we may find reason to believe them genuine hallucinations.

When a patient with peritonitis declares that a church congress is being held inside of her and says that she can feel it in the abdomen, no one knows what a congress in such a locality would feel like, but the patient mistakes it for a sense impression and hence it is an hallucination. Should the patient imagine that she hears the speeches of the contending parties in the congress, then of course there would be an auditory hallucination.

A so-called census of hallucinations has been made at the instigation of the Society for Psychical Research which is really a list and description of hallucinations which have occurred in recent times to such people as the promoters of the enterprise could induce to tell of them. It is probable that there is no person who has not frequently experienced them. Many of these are now on record, constituting quite a body of hallucinations. The purpose for which these records were made seems to have been the desire to prove that hallucinations are often veridical and hence give evidence of some unknown or hitherto unrecognized method of communicating ideas, except in folklore, when such communications are attributed to the interference of disembodied spirits in the affairs of mankind or an extra sense called telepathy by an organ not yet discovered. Those who believe in ghostly manifestations will find abundant evidence of them here, while those who believe in telepathy will gain confirmation of their doctrines. In the meantime those who still hold them to be hallucinations or specters will explain them as psychologic errors.

Parish in his work on *Hallucinations and Illusions* considers those of the S. P. R. catalogue with others which have been recorded by medical experts or derived from general literature. He endeavors to show that all hallucinations and illusions are phenomena of dissociation. Dissociation is manifestly abnormal association, and association is about synonymous with conception as we have used the terms.

When awake we may have hallucinations whenever our nerves are unduly excited or when we are in any abnormal condition, as from fatigue.

Hallucinations are a constant phenomenon of ecstasy, where they arise out of one-sided mental activity and intense concentration on single groups of ideas, conjoined with lowered sensibility. The best known cases are those of religious ecstasy, but religious ideas do not invariably furnish the material for "ecstatic vision." Philosophers, artists, and others whose habit of mind tends to deepen certain channels of thought, are also liable to such visitations. Any and every object of longing or desire, no matter how trivial, grotesque, or perverse, may become the object of ecstasy.—(P. 38.)

Emanuel Swedenborg was privileged to behold God himself. Engelbrecht relates how he was carried by the Holy Spirit through space to the gates of hell, and then borne in a golden chariot up into heaven, where he saw choirs of saints and angels singing round the throne, and received a message from God, delivered to him by an angel.—(P. 39.)

The multitudinous hallucinations recorded in history, like that of the demon of Socrates and those referred to in the former part of this chapter, are probably all hallucinations of ecstasy. Hallucinations are fundamentally classed by the sense deceived. Thus we have gustatory, tactual, motor, auditory, and visual hallucinations. Of gustatory and olfactory hallucinations, Parish says:

> Where hallucinations of taste have been noted they are mostly nauseous or poisonous (arsenic, copper, filth), and frequently give rise to refusal of nourishment, or it may be to continued spitting. In the early stages of paralysis, on the other hand, gustatory hallucinations of an agreeable nature are sometimes reported, the patient perhaps describing the enjoyment of all the various dishes of an imaginary menu. Olfactory hallucinations are, on the whole, infrequent, and are seldom of an agreeable character. The experiences of the patient who declared he smelt all the perfumes of Arabia and the East are exceptional, for hallucinations of this sense are, generally speaking, associated with delusions about bodily foulness, and odors of corruption and corpses, due to visceral disturbances. Lélut reports the case of an insane woman who declared that the pestilential odors she perceived arose from corpses buried in certain vaults under the Salpêtrière. Sometimes, haunted by the fear of being murdered, the sufferer perceives everywhere the fumes of charcoal, noxious gases, and particles of poisonous dust. Olfactory hallucinations seldom appear alone, but are generally associated with other sensory fallacies. Some authors consider that they belong more to the early stages of insanity. They are frequently found in association with local disease of the ovaries, and of the reproductive organs in general.—(Pp. 28, 29.)

Fallacies of touch seem usually to be represented by hallucination of external bodies crawling on the skin when in fact no such bodies exist. Hallucinations of insects, mice, and snakes are frequent.

There is not much to note concerning hallucinations of the tactile sense. . . .
It is only when a darkened intelligence "seizes upon them as a basis for a new conception of the ego and the environment," that they become of primary significance. But such significance may always be attributed to an hallucination of either of the higher senses, though opinion is divided as to which of these two senses plays the greater part.—(Pp. 29, 30.)

Hallucinations of pressure are more common than those of touch. In the dream state the walls of the building of a room may seem to contract until the sleeper is in a nightmare of trouble with the compression. These hallucinations are also common in certain diseased conditions.

Hallucinations of audition are very commonly caused by inflammation of the inner ear.

The sufferer hears taunting or insulting voices calling after him in the street, and making injurious insinuations about him, or sometimes unseen speakers incidentally let fall words which confirm his forebodings.—(P. 23.)
A kind of auditory hallucination worthy of special note is "audible thinking," wherein the patient hears his own thoughts spoken aloud, and imagines that they can be heard by everybody, or else hears them repeated or dictated to him by an imaginary being. Fallacious perceptions of the other senses are also not uncommon. Many sufferers see the persecutors who torment them from a distance by means of magnetic and electrical apparatus. They entertain kings and princesses, and receive angels' visits; all these hallucinations occur in a state of full consciousness.—(P. 24.)
Gall relates the case of a minister of state who constantly heard insulting words whispered into his left ear; and in the more recent literature of the subject such examples are no longer rare. According to Krafft-Ebing, the unilateral voices

are heard better when the other ear is closed—when, for instance, the patient is lying on it.—(P. 32.)

While walking alone she hears a voice calling her, she turns round, there is no one. While she is at her work familiar voices speak in her ear. She hears them on both sides, but chiefly on the right.—(P. 35.)

Hallucinations are . . . a frequent cause of violent and criminal acts; for instance, in hallucinatory insanity, epilepsy, hysteria, and somnambulism, and especially in delirious states (alcohol, morphia, cocaine, and typhus-delirium). Thrown into a paroxysm of terror by the phantoms which threaten him, or obsessed by his "voices," the sufferer snatches up a weapon and perhaps commits a murder or sets fire to the house. Or, again, despairing of escape from the enemies who pursue and mock him, he puts an end to his sufferings and his life at the same time, and often in a skilful and cunningly planned manner.—(P. 34.)

Tactual, auditory, and visual hallucinations most frequently occur on the hemianesthetic side.

Hallucinations of vision are more common than those of any other sense.

Thus Herr Von M—— told me that when taking his usual afternoon walk he used to see regularly on reaching a certain spot the head of the squadron returning from their daily exercise, and crossing the street at some little distance in front of him. One day when he had seen this as usual it occurred to him to wonder why the rest of the troops did not follow, and he soon discovered that the cavalry he had seen on this occasion were phantoms.—(P. 190.)

Some years ago, a friend and I rode—he on a bicycle, I on a tricycle—on an unusually dark night in summer from Glendalough to Rathdrum. It was drizzling rain, we had no lamps, and the road was overshadowed by trees on both sides, between which we could just see the sky-line. I was riding slowly and carefully some ten or twenty yards in advance, guiding myself by the sky-line, when my machine chanced to pass over a piece of tin or something else in the road that made a great crash. Presently my companion came up, calling to me in great concern. He had seen through the gloom my machine upset and me flung from it.—(Pp. 191, 192.)

Gregory mentions the case of a patient in whom the seizure was always preceded by the apparition of a hideous old woman in a red cloak, who advanced and struck him on the head with her cane, whereupon he fell to the ground in convulsions. In another case the devil appeared in a shadowy form. Sometimes the apparitions are less frightful. Conolly tells of a patient who saw, in the last few moments before loss of consciousness, pleasant landscapes spread out before him.—(P. 33.)

For example, the commonest visual hallucinations (in which black and red play a leading part) are black rats, cats, snakes, and spiders, shining stars, fiery spheres, and so on. But these do not remain motionless. Either they go diagonally across the patient's field of vision, in which case they proceed from the hemianæsthetic side; or else (generally) they come from behind the patient, hasten past, and disappear in the distance. In this case also the apparitions occur on the hemianæsthetic side. . . . These premonitory hallucinations haunt the sufferer even by day, but in the night they become much more persistent and vivid, and what was only a passing vision before, develops into a long scene, in which the patient is called upon to take a part.—(P. 35.)

Sufficient illustrations have perhaps been given to exhibit the fundamental classification of hallucinations. Were I writing a treatise on hallucinations rather than a condensed account of the subject, every class should be sub-classified by the agency through which they are produced. This classification would give us, (1) the hallucinations of dreams, (2) the hallucinations caused by subverted sensation or ecstasy, under which are included the phenomena of crystal vision, (3) the hallucinations of suggestion or hypnotism, (4) the hallucinations of intoxicants, (5) the hallucinations of disease.

In sleep the senses are dormant while the functions of life continue. Sense impressions are only instantaneous, but feeling impressions endure as long as the cause acts, although they may become dulled by repetition or unrecognized by habit. It is well known that a sense impression may give rise to a feeling if it is too intense. It is an old doctrine of psychology that sensation is inversely proportional to feeling, and it remains true to this extent, that a sense impression may be neglected, that is, we may not consider the cause though we may consider the effect, when the impression will give rise to a feeling. In the dream state sensation lies dormant and feeling has the psychic field to itself.

In sleep sense impressions frequently impinge upon the organs: lights appear in the darkened room, sounds are made which produce some slight effect upon the ear, and to the sleeping person there come many tactual impressions, all of which are interpreted as feelings and produce hallucinations because feelings are so intimately associated with external objects; these are feeling hallucinations.

On the other hand if on a cold night the clothing is partially removed from the body the feeling of discomfort is quite likely to produce an hallucination. Drops of water falling upon the face of the sleeper may have the same effect.

The bedcover pressing on the arm is embraced as a mistress or felt as a heavy weight; a dream of being impaled, that is to say, of standing on a stake, the point of which was thrust through the foot, has been known to arise from the pressure of a straw lodged between the toes; a covering which has slipped to the ground is sometimes a source of great embarrassment, when it causes us to dream of appearing half clad in the street or at a social gathering; or it may call up visions of skating, Alpine travels, Polar expeditions, and these again may suddenly end in the feeling of falling into a gulf, due to a slight alteration of the sleeper's position in bed. Gregory, when he had a hot-water bottle at his feet, dreamed that he was climbing Etna and walking on hot lava. Purkinje says: "If our hand has become numb by pressure, in the dream-state it may appear as something strange and gruesome touching us, and if the whole side is affected, we imagine that a strange bedfellow, whom we cannot get rid of, is stretched beside us.—(Pp. 54, 55.)

The influence of position during sleep is generally exhibited in one of the following ways: (1) The position of a member may be perceived more or less correctly, but suggest an attitude; for instance, if the foot is stretched and bent back it suggests the dream of standing on tip-toe to reach something; (2) the strained position may be taken to be part of a movement and the dreamer seem to be dancing on his toes; (3) the movements may appear to be executed by some one else; (4) sometimes the movements seem to be impeded; (5) the affected member may be changed in the dream into some animal or inanimate object of analogous form; (6) sometimes the dream-perception of the member gives rise to abstract ideas, which it symbolizes; for instance, the perception of several fingers may give rise to dreams of numbers and calculations. —(P. 55.)

A mustard plaster on the head may cause a man to dream of an Indian conflict in which he is scalped, as I have observed.

Thus Herrmann, when suffering from an attack of colic, dreamed that his abdomen was opened, and an operation performed on the sympathetic nerve. Others dream of going up for examinations. The house-wife dreams she is giving a party, and that all her dainties are burnt up, and so on.— (P. 56.)

An individual directed his servant to sprinkle his pillow sometimes after he was asleep (leaving the choice of the particular night to the servant) with a perfume which he had only used during a certain stay in the country, but to which he had then taken a great fancy. On those nights he visited again in his dreams the scenes associated in his mind with the perfume. The occurrence of imaginary tastes and smells in dreams is very rare, so much so that it has been altogether denied by many observers. Still a few cases have been reported.— (P. 54.)

Hallucinations of ecstasy often arise with persons engaged in profound abstract thought. Philosophers, poets, literary men, generals, and divines are peculiarly subject to them. Extreme ethical emotions are apt in begetting hallucinations. It is through all of these cases that the world's literature is replete with accounts of hallucinations. Perhaps every great man has had them.

We have abundantly affirmed and illustrated the doctrine that sense impressions are instantaneous, and the judgments which we form from sense impressions are instantaneous, while feeling impressions endure while the cause acts. It is possible for us to concentrate the attention upon the impressions received by one organ, but if we fixate the attention on an interrupted succession of like impressions we overthrow or subvert judgment. As we must at every instant go on to form a new judgment, the supposed concentration of attention sets the mind adrift to follow feeling impressions wherever they may lead. This subverted sensation I call ecstasy.

We make a multitude of judgments of recognition at one glance of the eye about the room which we occupy, or over the landscape when we are out of doors. Now, if we can fixate the attention of the eye or the ear and abstract the mind from all other sense impressions, hallucinations may be produced. This secret has been an open one to those who have practiced divination in the departed centuries. There is a vast body of literature on the subject, though it relates chiefly to the abstraction of vision.

Even as I write, the boys on the street are crying the New York papers and tempting purchasers with stories of divination by crystal vision.

In crystal vision the percipient attempts to occupy his mind in the contemplation of a constantly renewed sense impression, while the mind in fact is recalling concepts from memory which he ascribes to hallucinatory objects in the glass; that is, he forms judgments of things not seen but remembered by suggestion from feeling impressions. We may express this idea in still another way. In crystal-vision experiments the mind of the percipient is engaged in recalling memories which may be determined by the feelings or may arise at random, for it is impossible for the waking mind to cease operations. As the thing expected or looked for in the glass does not appear, these memory images are projected into the glass.

> The percipient strives to banish all conscious thought from his mind, and fixes his gaze continuously on a "Braid's crystal," a burning glass in a dark frame, a glass of water or some similar reflecting object. Many persons after gazing thus for some time begin to see pictures in the crystal, the spire of the parish church perhaps, or familiar faces.—(P. 63.)

An eye-witness relates the following anecdote of an occurrence in Egypt:

> His curiosity was excited by Mr. Salt, the English Consul-General, who, on suspecting his servants of theft, sent for a magician. Mr. Salt himself selected a boy as seer, while the magician occupied himself with writing charms on pieces of paper which, with incense and perfumes, were afterwards burned in a brazier of charcoal; then, drawing a diagram in the boy's right palm, into the middle of which he poured some

ink, he bade him look fixedly into it. After various visions had come and gone, the form of the guilty person appeared to the boy, and was recognized by the description he gave. On being arrested, the thief thus strangely convicted confessed his crime.—(P. 64.)

Just as visual images may be called up by gazing on a shining object, so by placing a sea-shell to the ear it is possible to induce auditory hallucinations. I therefore class such hallucinations with crystal visions, which they resemble in their content. This analogy is borne out by cases like that of the lady who, if she listened to the shell after a dinner-party, generally heard repeated, not the conversation of her "lawful interlocutor" to which her attention had been directed, but the talk of her neighbors on the other side, which she had not consciously noted at the time.—(P. 70.)

All modes of ecstatic hallucination are of this character. It is the abstraction of attention to the particular object while waiting for a judgment of cognition or recognition to come through the intellectual faculties, while instantaneous judgments continue to be made through the emotional faculties. The consideration of this fact leads us to restate that which may seem already to have been abundantly affirmed, that the vital organs of metabolism, circulation, motility, and reproduction are the end organs of feeling, while in the nervous system we find organs of feeling and intellection.

The third class of hallucinations comes from the land of suggestion. Much of the intellectual activity of mankind is acception, or the receiving of judgments made by others through the agency of speech; words are heard or seen that express judgments which we accept as valid. So much of intellectual life is of this character that we are trained in the ability of acception. This ability runs astray with some persons because there goes not with it the habit of constant verification. The speech of human beings must be verified in the same manner that natural language in presentation and representation must be verified. He who accepts the judgments of others without intellectual verification is eminently qualified for hypnotic suggestion.

There are some people so naïve in their interpretation of expressed judgments as to suppose that what is told them must be either truth or falsehood, not being able to distinguish a fallacy from a lie. This simplicity in weighing the judgments of others is highly conducive to the development of hypnotic intellects.

Frau U., an innkeeper's wife, forty-five years of age, an extremely suggestible subject (so much so that while awake a mere assurance that she could not move her limbs deprived her of all power of movement), was hypnotized by me, and the post-hypnotic suggestion given that each time A., who was present, should cough, a fly would alight on her brow. The hallucination was realized; at each cough of A.'s she raised her hand to her forehead and looked up into the air as though watching a fly. This did not prevent her, however, from continuing with animation her conversation with me on the preparations for her daughter's approaching marriage. Her prompt reaction to suggestions given in ordinary life rendered her post-hypnotic suggestibility valueless as a test of her state of consciousness.

Bernheim communicates the following case of a young girl, of unusual intelligence, and free from hysterical tendency: I arranged that on waking she should see an imaginary rose. She saw it, touched and smelt it, and described it to me; but knowing that I might have given her a suggestion, she asked me if the rose was a real or imaginary one, adding that it was quite impossible for her to tell the difference. I told her that it was imaginary. She believed me, and yet found that by no effort of the will could she make it disappear. "I can still see and touch it," she said, "as though it were natural; and if you were to show me a real rose beside it, or instead of it, I should not be able to tell the one from the other." All this time she was thoroughly awake, and talked quietly with me about the apparition.—(P. 62.)

In a former chapter it was stated that the corpuscles of the blood are unicellular organisms and that the red corpuscles are built into the system, so that every part is composed of unicellular organisms. Each of these organisms is endowed with the rudiments of life and mind which they take with them into the human system. The phenomena of hypnotism reënforce the discoveries of physiology and confirm the doctrine that the entire body is the seat of consciousness and that the nervous system constitutes the special apparatus of inference. This leads us to a theory of multiple seats of consciousness which is demonstrated by the phenomena of hypnotism, a tempting subject which we are compelled to ignore by reason of the limitations of our argument.

Hallucinations caused by intoxicants are well known. Those occurring through the immoderate consumption of alcoholic drinks are most common.

. . . The hallucinations . . . are generally of a depressing nature, and terrifying impressions predominate. True, sweet voices are sometimes heard, melodies delight the ear, and fair landscapes appear before the eyes, but this seldom lasts long, monsters and serpents take the place of flowers, and the visions shift about and are mingled together. Vermin, reptiles, etc., appear in great numbers, such for instance as the rats, cats, snakes, mice, and monkeys, which fill the visions of delirium tremens. Thus Brierre de Boismont found among twenty-one cases—three of them severe—twenty in which hallucinations of vermin and such creatures were seen swarming over the bed and up the walls. Other sensory delusions of a purely fantastic nature are not lacking. Sometimes black men appear who grimace and threaten, then climb the walls, or vanish up the chimney. In other cases the visions arise out of the daily occupations of the patient, or out of his past experience.—(Pp. 41, 42.)

In addition to alcoholic beverages many drugs produce hallucinations, as opium, hashish, santonin, etc. Among the tribes of the western plains of the United States a cactus known as peyote is widely used in their religious rites. The plants themselves, when made into decoctions or when eaten as dried fruit, produce a variety of effects, among which are those of color vision. Dr. Theodate Smith, an expert in experimental psychology, has furnished me with the following memoranda of an experiment on herself in the use of the peyote. Earlier trials produced in part very disagreeable effects and in part excessive motor excitement, but after repeated trials color visions came only when she placed herself under some restraint from motor activity; then there appeared a set of retinal effects in a succession of dissolving views which she described to an attendant who was charged with making a record of her words.

The following is an extract from this record:

Branches of coral, in color a deep, beautiful blue.
Flattened forms of coral shape, deep purple changing to red with ruby red tips.
An electric fountain, many colors.
Colors of a peacock's tail, form somewhat indistinct.
Flashes of light over the whole retinal field; predominant color a wonderful intense green.
Flower forms—quantities of violets, yellow in color, flickering light over them, also yellow.
Deep opal-blue rings running outward from a center and in constant motion.

Beautiful green light, like light in an electric fountain; no special form.

A complex Grecian pattern, deep blue with white dots suggesting snowflakes over it.

This changes through many tints of blue to turquoise blue; the form becomes a bowl and pitcher ornamented with gold.

A ship with square sails on the *bluest* ocean, intensely blue.

Blue aureoles encircling everything as I half open my eyes in dim light.

Strings of beads of many colors.

Embroidered leather with rainbow colors flickering over it as if from a stained-glass window.

Nine leaves of silvery gray conventionalized.

Cat's fur, but colored blue and white.

The blue becomes lines and forms, the outline of a big centipede.

Venetian glass, amethyst tinted, shades from light to dark, wavy lines running through it, forms not distinct.

An escutcheon, quarterings of blue, steely blue, a shield with lines; around the shield four swallow tails. These enlarge, cover and finally blot out the shield.

A shining laurel leaf.

A beautiful chandelier, richly jeweled and blazing with light.

A stained-glass window, red, blue, and amber, colors rich and deep, forms not well defined.

A crazy quilt, pretty but very crazy. A transparent flexible lily shape, with wavy lines running through it like bird-of-paradise feathers; no color in the form itself, but it seems to float in the midst of colored light.

Phosphorescent fishes' eyes.

Fish scales of wonderful green, changing to shell shapes in the green light.

A picture of an arctic sunset, with silver rays rising from it, and far off on the edge an aureole of beautiful blue.

A ceiling from which hang ribbon cards of every color.

A camel with gorgeous trappings, with a palm tree behind him.

Embroidery of red chrysanthemums, variously mixed with pale pinks and yellows.

All of the North American tribes have intoxicants that produce hallucinations, but they supplement these intoxicants with many rites such as dancing, singing, ululation, the beating of drums, and the tormenting of the body by various painful operations, all designed to produce ecstatic states and the consequent hallucinations.

Among all tribal men many hallucinations are supposed to be veridical, as some are supposed to be by certain members of the Society for Psychical Research. So tribesmen resort to the agencies which produce both hallucinations and illusions to obtain a view of the world about them, of the past and of the future, in order that their conduct may be governed by this superior knowledge.

Had our psychologists attempted to make a "census of waking hallucinations in the sane" among the North American Indians they would have found a hundred per cent. ready to testify in their favor. It is the universal belief in savagery, for in that stage of culture all men produce hallucinations for divination—for which times and seasons are regularly appointed and systematic means employed. But the savage always recognizes that some visions are not veridical. False spirits may have testified or some evil being may by black art have vitiated the ceremony or the percipient may have been unable to properly read the communication, for communications are told in ambiguous terms. It is very interesting to read these communications recorded in the annals of the Society, for we find that after all it is often necessary to wait for a time to discover an event which will fit the hallucinations.

With the hallucinations already considered, those appearing in the course of acute somatic diseases, and as a result of them, seem naturally to be classed. Here, as in the delirious states associated with intoxication, the swarming of the hallucinations is characteristic. This resemblance is not accidental. Indeed, the delirious states of somatic disease may, in part at least, be referred to intoxication. But of no less importance are the rise of temperature, acceleration of metabolic processes, and disturbances of circulation in the brain cavity (first, active hyperæmia; later, in enfeebled action of the heart, venous stasis), the importance of which is indicated in typhus, for instance, by the parallelism between the violence of the delirium and the temperature curve. The initial hallucinatory visions of typhus, smallpox, and intermittent fever, occurring before the other causes have had time to act, are on the other hand to be attributed to the direct influence of the specific virus of the fever, as also the afebrile delusions, sometimes occurring in intermittent fever in place of the fever attack, and the visual and auditory hallucinations which are observed in smallpox between the eruptive fever and the fever of the suppurating stage.

Hallucinations also occur in the decline of the disease, during the period of convalescence. First they appear singly, in association with those of the fever, and are often recognized by the patient as such and concealed from those around him. But soon they overmaster the sufferer, and delirious states are developed, or states resembling hallucinatory insanity, in which visions of corpses, death's-heads, mocking voices, and offensive olfactory and gustatory hallucinations play a part. Of an equally distressing nature are most of the sensory fallacies of collapse-delirium, and those which sometimes precede death. In tuberculosis, on the other hand, they are often of an agreeable nature, corresponding to the euphoria which is so characteristic of this disease.—(Pp. 48-50.)

The most frequently quoted of all sense-deceptions are those of insanity. Some authors have sought to divide them according to their origin into "idiopathic," those which are primary but which may also occur in secondary consensual morbid states, and "symptomatic," those which occur only as a secondary symptom of insanity. In any case a distinction ought to be drawn between sporadic hallucinations not associated with particular emotional states, and hallucinations which reflect the ruling mental tone. This distinction has prognostic importance, since observation seems to prove that hallucinations depending on certain morbid emotional states are capable of disappearing with them, whilst independent hallucinations seldom admit of cure, and pass over into the state of secondary psychical weakness.

The particular forms of insanity in which hallucinations most frequently occur are such as are associated with dream-like beclouding of the intellect. Thus they are a frequent phenomenon of amentia, but are seldom seen in acute dementia with its deep-reaching paralysis of the higher psychical functions. Opinion as to the frequency of sensory hallucinations in melancholia has altered very much of late years, chiefly because of the altered meaning of the term, and because cases previously classed under melancholia are now referred to other groups. Thus, while hallucinations were at one time regarded as frequent phenomena of this state, they are now held to be rare, or altogether absent from it. In mania hallucinations only appear when there is clouding of consciousness, and are generally vague and indistinct. On the other hand, illusions are frequent, and mistakes of identity are specially characteristic of this state, though not absent from other forms of insanity. Snell, who devotes an article to them, is of opinion that the confusions are not so much caused by mere resemblance, but that a general psychological law lies at their root; that the patient is powerless to escape from the familiar thought-channels, and therefore grafts his new impressions on to his old opinions and ideas. In *folie circulaire* hallucinations occur in the maniacal period in association with profound mental disturbance, but as regards their occurrence in the melancholic phase opinion is again divided.

Delusional insanity and Paranoia, on the other hand, abound in hallucinations, so much so that some forms classed under this head are designated "hallucinated insanity" (*hallucinatorischer Wahnsinn*), and "paranoia hallucinatoria." The sense-deceptions of delusional insanity are vivid in their externalization, and resemble in their content the fixed ideas which they embody. In cases which end in mental decay the hallucinations frequently persist long. In depressive monomania they are more fragmentary and vague, but are often kept alive by distressing dreams. . . . The sufferer hears taunting or insulting voices calling after him in the street, and making injurious insinuations about him, or sometimes unseen speakers incidentally let fall words which confirm his forebodings.—(Pp. 20-23.)

The physiological conception of memory is that concepts are impressed upon the brain and the nervous system as elements of structure. Memory is thus a function of structure. The revival of concepts is recollection; such revival is accomplished by a sense or feeling impression, but a sense or feeling impression is a force or mode of motion which is utilized by conditions so that the central consciousness or consciousness of the brain is subject to conditions which we call causation. Thought is therefore explained physiologically by the late discovery that sense and feeling impressions traverse paths along the fibrous nerves which are diverted by the ganglionic nerves to different tracts of the brain, where concepts are recorded as structural elements. Thus hallucinations are explained by referring them to the mechanism of the brain and showing how by such mechanism incongruous concepts may be aroused by defects in its working.

Now we are prepared to reaffirm that a judgment of sensation must be verified to become a cognition, for if a judgment of sensation is an hallucination there is no cognition. Many of our sensations may be verified by repetition, and it is often the case that this method establishes their verity.

The hallucination caused by subjective audition cannot be disproved by a repetition of the hallucination caused by an injury to the middle ear. An hallucination which is a color vision cannot be shown not to be veridical in this manner, for it may continue while the intoxication lasts. The ultimate test of the verity of a sensation is an appeal to a higher faculty of the mind, which is perception, that yet requires explication.

The person who had an hallucination of a church congress in her stomach was not in a condition to appeal to a higher faculty. Before she realizes that she has an hallucination her malady must be cured. The man who believes in ghosts when he has an hallucination of his dead child appearing to him in the cerements of the tomb can best be shown that it is an hallucination by curing the malady in his understanding.

CHAPTER XXI

FALLACIES OF PERCEPTION

We have found that sense impressions cause events of consciousness which produce judgments by recalling concepts of sensation, such concepts being reinforced and developed by the addition of new judgments. Judgments of perception still employ the same sense impressions in the construction of new concepts of form, while concepts of form are recalled when a judgment of form is made. A new concept of form is constituted by the increment of a new judgment of form. Therefore concepts of sensation are concepts of kind, while concepts of perception are concepts of form. As a judgment of sensation must always precede a judgment of perception, the same sense impression which gives rise to a judgment of sensation will, in the maturer mind of the infant, also give rise to a judgment of perception; therefore we are compelled to reconsider the sense impressions from which perceptions arise. Having already found how judgments of perception are considered and how such judgments are verified, we have now to exhibit in what manner there comes into existence a multitude of judgments of perception which are never verified, and yet are entertained in the mind as if they were veridical.

Fallacies of perceptions are errors of judgment respecting forms. Such judgments may occur through unverified judgments of sensation, and the fallacy is repeated in a higher state of mind. Judgments, when they are first made, are of slow growth, but when once made, by repetition they become habitual and do not arise in the corticle consciousness.

The human mind cannot perceive form without first sensing kind. On the other hand it seems almost impossible to sense a kind without at the same time perceiving a form, though we may pay attention to the kind or to the form at will. In our discussion of fallacies of sensation, we have tried to pay attention to the kind, but we have found that kinds were usually expressed as forms. The experimental observer, Miss Smith, not only spoke of colors as dissolving in succession, but at the same time the colors themselves were explained as forms. Most of the fallacies of sensation which we have cited in this discussion, most of those which appear in the general literature of the subject, and most of those which occur in experience are not only hallucinations of sensation, but they are also specters of perception, because the human mind rarely senses an object without at the same time perceiving the object. When I see the color of the rose, I see the rose as a form. When I see the color of the cloud, I see the cloud. When a word is pronounced in my hearing I hear the sound as a sound, perceive the person in the other room represented vicariously by the voice, and at the same time hear the word as a word and as a symbol of meaning. In general, the description of a sensation is best accomplished in terms of perception.

We must know things as kinds before we know them as forms, and we must first judge of things as kinds before we judge of them as forms. But when we already know things as kinds, we can re-cognize them as kinds by instantaneous judgments, and at once go on to cognize them as forms, or to make judgments about them as forms. In a former chapter, fallacies of sensation were often described in terms of perception, for they seem always to produce fallacies of perception, and in the state of mind under which they are produced it is the forms, not the kinds, which are of chief interest to the subject.

There are many misperceptions; so common are they as to be scarcely noticed. If a person will observe his own thoughts from moment to moment, he will be surprised at the number of fallacious perceptions which he makes, some of which are immediately corrected, others are corrected after lapse of time, and probably many others that are never corrected, because of their insignificance in the practical affairs of life. These errors of judgment are especially common in audition and vision, the two senses most highly vicarious. A sound may be obscure by reason of its faintness, or by reason of diverted attention. Sight may be obscure by reason of the twilight, or it may be obscure because attention is elsewhere directed. All such impressions may be veridical or may be fallacious. If I am intently listening for a sound I may interpret a sight for a sound; if I am intently looking for an object, I may interpret a sound for a sight. If I am intently listening for a particular sound and hear another, I may interpret it for the one I was expecting; if I am intently gazing in expectation of seeing one object, and another falls upon the field of vision, I may see in it the one for which I was intently gazing. These are all misperceptions.

I draw nine black lines on white paper, as shown in Fig. 1, and you see them as lines on paper. Now close one eye, and lift the page horizontally nearly to the height of the eye, and these lines will appear as pins. By a little manipulation you can see them now as pins and now as lines. You know they are not pins, yet you see them as pins;

Fig. 1.

that is, you have formed a habit of interpreting sense impressions like those made by the lines when they are in certain attitudes as marks or symbols of standing objects set as pins, stakes, men, or trees, and so thoroughly established is this habit that such an attitude of lines may be interpreted as standing objects when they are not,

Fig. 2.

and you will affirm that they are lines at one time and standing objects at another. This is one of the standard illustrations of misperception. Now will be understood the statement when it is affirmed that only color is manifested to the eye by the object, and that when such a judgment is formed it may or may not be valid, but that the color is interpreted as a symbol of the object in a judgment of perception.

Before me as I write there is a steam register, which is covered with a tablet composed of bars with interspaces, the bars being arranged in patterns; a drawing of a portion of this tablet is illustrated in the accompanying diagram Fig. 2.

Looking upon it in the ordinary position in which a book is read it appears as a pattern of bars; turn the top of the book to the left in such a manner as to see the bars obliquely, and it appears as a collection of crates or boxes inclined one upon another; turn it again so that the direction of sight is changed ninety degrees from the first position, and you can see it as a series of steps like a stairway, every tread having a series of reëntrant angles. Again, we see that in vision nothing but color as in a flat is given to consciousness, and that form comes by interpretation or inference. Deftness in inference is acquired by practice; that is, it is the result of experience. We come to interpret lines in this manner as meaning form by the experience of every moment of waking life, and inherit the skill from a long line of ancestors, so that our powers of perceiving formed in this manner are both inherited and habitual, or, as I prefer to say, both instinctive and habitual, and that which is both inherited and habitual is intuitive.

Light and shade are interpreted as deftly as lines, and we can see forms without other colors, so that a portrait which you know is only light and shade, is a symbol of the form and expression of a human face. But there are other colors both in nature and in art, and we instinctively and habitually interpret all colors as forms; but sometimes we see colors without seeing forms. The illusions of inference by the interpretation of lines in vision have been the subject of much investigation in psycho-physics, which is one branch of scientific psychology. But adequate experiments have not yet been made in light and shade, and in other colors when not represented by lines. The doctrine dates back to the days of Berkeley, who set forth the nature of perception in vision in such manner that it has become a classic, though he afterward devoted his energies to the propagation of fallacies in metaphysics and tar-water.

From time to time during the last thirty years, I have studied the nature of perception in myself and in others. Especially have I studied it as a mental phenomenon in the untutored Indians of North America. On every hand these facts have appeared: first, that every perception as a judgment involves an interpretation; second, that perceptions may be true or erroneous, as inferences are valid or invalid; and third, that visual perception itself is acquired by experience.

Among the Indians, I have found that at first lines are not easily interpreted, so that pictures in lines do not seem to represent forms; but the power of interpreting forms by lines is rapidly gained. I have found also that the power of interpreting light and shade is great in the savage for natural objects, but must be cultivated for unknown objects of art. And, again, I have found that the power of interpreting the miscellaneous colors of pictures is well developed when they represent things with which they are already familiar, but that it is necessary to familiarize them with things to develop the power of interpreting unknown forms.

Again, in topographic maps, relief is represented usually by light and shade in hachures, but in the best maps relief is represented by lines which follow the contour at equal intervals of altitude. Such maps cannot be read by the inexperienced man, but he can develop the power so that a contour map will seem to be a picture of mountains and valleys and of hills and dales. Experience has taught me that this power is more easily gained and greatly assisted by representing relief in one color and drainage in another, as in blue; for when the knowledge that water is blue is represented in the map as blue, it will carry the streams down and aid in the perception of the relief.

From the illustrations which have been given it will perhaps be made clear that perception is the interpretation of a symbol, and that the power of interpretation comes by experience. We are constantly perceiving with all our senses, but sounds and sights are the most abundant, coming in hosts with every minute of wakefulness, and a habit of interpretation is formed which is conjoined with an inherited aptness. External forms do not come to the eye or the ear as consciousness, but only to the mind as inferences. Habitual judgments of the mind which are illusions because unverified, may occur again and again in millions of cases, and the repetition but confirms the illusion, and such intuitive illusions can hardly be dispelled even by overwhelming knowledge, but the truth and the error will appear side by side and be entertained as verities, and the mind will search for some metaphysical explanation of them. As a last resort of logic, it will assume the existence of a mystery, and be confirmed in the doctrine that the universe is contradictory.

Our forefathers called the sky a firmament. It was believed to be a solid which presented a surface toward us, and this misconception is universal among barbaric and savage people. By the Indian the sky is supposed to be ice, or some other crystalline solid, and it does appear to be a surface, in spite of our knowing that it is not. This arises from the fact that we always discover color on surfaces, and when surfaces are removed usually colors are changed. We have thus as individuals and as a race in all generations habitually considered color to be a symbol of surface. That which is habit in the interpretation of a sense impression contradicts that which we have learned by various operations of reasoning from other sense data. Thus habitual illusions often contradict certitudes, as they may be discovered by the higher forms of reason, and we often entertain certitudes and fallacies as if co-existent, and the world seems to be contradictory. These judgments have a curious effect on the mind, for the contradictory judgments may both be held in a vague way to be certitudes and still in a vague way to be fallacies, until finally this is explained by a theory, that both are unknown and unknowable noumena which are manifested by deceptive phenomena. So habits of judgment are formed which are difficult to eradicate.

To unverified perception the rainbow as a form with a surface has been established, because of the habit of interpreting color as a mark of surface; this fallacy is common, perhaps universal. The clouds often seem to be painted upon the sky, or to be moving along the sky, but the trained meteorologist in time learns to distinguish clouds as forms, and discovers fleeting figures in them, and he still further discovers the relative position of clouds by recognizing the near from the far, and yet, to the untrained observer, there still lingers an element of fallacy.

It was long believed that the earth has ends, corners, foundation, and a flat upper surface. When it was discovered that the earth is a spheroid, the illusion of up and down as components of direction at right angles to a flat plane was dispelled, and a concept substituted of down toward the center and up from the center. While a few grasped the idea, the many still held to the old, and now, after more than two thousand years, there are people who have not mastered the concept.

One man sees the disc of the moon when it is riding high as having the size of the top of a teacup, another as large as a cartwheel. But the moon will seem to be larger than a barn if it is seen behind a distant barn, or it may seem to be as large as a great mountain when it rises behind such mountain, and yet every intelligent man knows the moon to be 2,162 miles in diameter. As the moon rides the heavens, it seems to be this side of the surface of the sky, although we know that there is no such surface.

Such habitual judgments of space and form seem to contradict our knowledge. When knowledge contradicts primitive and habitual judgments, there is a pseudo-belief in both, and the universe seems contradictory.

The sun appears to us as a mile or two away, but we know that it is ninety-three millions of miles away. The sun seems very much nearer to us when it rides high in the heavens than when it comes up behind a near hill, or when it rises behind a distant mountain with intervening plains. What we know and what appears seem to contradict each other; and antinomies are invented to explain these contradictions.

By a natural process of fallacious judgment, the idea of space as void is developed as an existent thing or body. This is the ghost of space—the creation of an entity out of nothing. I may remove the furniture from the room, it is still filled with air; I may remove the air from the room, it is still filled with ether. We may suppose it possible to remove the ether, then nothing—void—remains, but man has no means by which to accomplish the feat, and we call the air and the ether space. The space of which we speak is occupied; it is the space inclosed by the walls, occupied by air and ether. We may measure its dimensions by measuring the walls, but we cannot measure the void. We can by no possibility consider non-space or void as a term of reality; we can consider only the walls as the real terms. If we reason about it mathematically and call it x, the meaning of the x in the equation is finally resolved by expressing it in terms of body as they are represented by surface. This non-space has no number; it is not one or many in one—it is nothing. It is not extension as figure or structure—it is nothing. Void space should be called voidable space, as voidable by one set of extensions when filled by another. The fallacy concerning space is born of careless naming. No harm is done by this popular misperception of space until we use it in reasoning as a term of reality; then the attributes of space may be anything because they are nothing. Such space is the occult noumenon, the reified void. This is the space of Kant, and usually the space of metaphysic. It is the reification of "pure" property, void of all extension which can have no relations; that which is without relation is non-existent.

When I consider the distance from here to San Francisco, I may think of the plateaus, mountains, hills, and valleys which have to be surmounted and crossed in traversing the distance, or I may think of the days required to make the journey. Yet I imply or posit the plateaus, mountains, hills, and valleys; so when I consider the distance to the sun I posit the spacial particles which intervene, though I may cancel their consideration, but if I affirm that space as nothing intervenes I affirm a fallacy. By calling it a five days' journey I do not annihilate the topography.

In the earlier stages of culture, when there was no knowledge of air and ether, this was the judgment of mankind, but I must not go on repeating this judgment when I know the truth. If the primeval judgments are held to be veridical, and scientific judgments also to be veridical, then the world is contradictory. Metaphysicians formulate these erroneous judgments and scientific judgments as antinomies.

Misperceptions have been discussed sufficiently for present purposes as exhibiting the characteristics of illusions. I go on to discuss specters which are derived from hallucinations in order to set forth the characteristics of delusions.

It will not be necessary for us to rediscuss all the hallucinations set forth in the last chapter, but it may be well to recall some of them as illustrating these principles.

Fallacies of sensation in the metabolic sense seem rarely to produce fallacies of perception. If they do arise they are vague. It is rarely, indeed, when they are produced that the deceived mind refers them to distinct objects as forms, but in extreme cases deceptive forms appear, especially in the case of odors, as when the subject refers such odors to the bodies of the dead, as the woman who referred the pestilential odors which she believed she sensed to the corpses buried under the Salpêtrière.

Usually the fallacies of touch produce illusions which the deceived subject attributes to some form of object which touches the skin; commonly these objects are insects.

In my study of the literature of hallucinations, I find but few hallucinations of the sense of pressure; yet there are a few, as when people dream or insanely imagine that they are enclosed by walls which are ever becoming narrower and thus compressing them.

To the person who has all of the senses, most of the hallucinations occur in audition and vision, because of the function which spoken and written language performs in the ideation of these senses.

Hallucinatory sounds often produce phantasmal words spoken by spectral persons.

The spectral person may be the self, or it may be another or a congress of others. When the voices of others are falsely perceived as persons, these others are specters.

Specters may be classified by senses deceived, and subclassified by the agencies through which they are produced. The class of specters derived from hallucinations of vision we will treat as thus subclassified, for the purpose of illustrating the doctrine.

When the nervous system is relaxed in slumber so that sense impressions carried by the fibrous nerves are directed by the ganglionic nerves at random to different portions of the cortex of the brain, sense impressions are produced upon that organ which result in dreams, and the imagination of the sleeper revels in wonderland. As these are of nightly occurrence, and all men dream, the ghosts of dreamland that fill the sleeping life are remembered in many a revery of the waking life.

In the culture reached at the stage of tribal society, images reflected by the water or other shining objects are supposed to be ghosts. Echoes are also referred to ghosts. Thus there is an explanation given to the common phenomena of reflected sights and sounds by attributing them to the ghosts which appear in dreams.

Hallucinations of ecstasy always seem to produce phantasms or specters of vision. Hence the specters seen by the great men of the world who have had a weight of affairs to contemplate—too great for their mental faculties; hence the specters seen by divines and poets. Such ghosts can be summoned readily by those phenomena which we have classified under the general designation of crystal vision, for the mind seems able by an effort of will to abstract attention from sense impressions in a fixed gaze upon a bright object, and then to be deluded with false judgments about such bright objects, seeing in the bright object itself many strange forms which are recalled from memory and projected into many incongruous relations of space. The phantastic images of the Braid's crystal are thus ghosts summoned from the vasty deep of hallucination.

The hallucinations of hypnotism make men see things which do not exist, and prohibit men from seeing things upon which their eyes are turned, when the patient is under the influence of the words or of the suggestions of a dominant operator.

Chloroform, ether, peyote, and many other drugs bring us hallucinations under conscious experimentation. But there are many intoxicants. In tribal society intoxicants are used for the purpose of producing hallucinations; in modern society alcohol is used as a beverage to produce gustatory pleasure; but in whatever way intoxicants are used hallucinations are produced. The hallucinations of obscure vision, reinforced by the hallucinations of dreaming, reinforced by the hallucinations of hypnotism, are still reinforced by the hallucinations of intoxication, until ghosts are the common property of mankind, and only through scientific training is the mind able to banish them. But these ghosts, while they affect the lives of many sane people, do not take entire possession of them.

When, however, the mind is diseased, the hallucinations of sane life take possession of the person.

The poor soul possessed by hallucination becomes a prey to melancholia, hysteria, and dementia. But the mind of the superstitious man, who is ever recalling the phantasms born of hallucination, is exploiting upon the brink of the sea of hallucination into which he may plunge by insanity. While ghosts may be smelled, touched, or heard, yet they are more commonly seen for vision is the most idealistic sense.

In the realm of ghosts there are five provinces—the land of dreams, the land of ecstasy, the land of suggestion, the land of intoxication, and the land of insanity. In tribal society ghosts of animals prevail, while in civilized society ghosts of men prevail. If you were talking to a savage about some unusual occurrence, he would tell you how he had been warned by a bear, that a hummingbird had appeared, that a rattlesnake had crossed his way, that an eagle came to him in his dreams. Homer's ghosts all appear as deities in the guise of human beings.

For twenty centuries metaphysic has been in search of the noumenon—the thing-in-itself. For a long time it spoke with disrespect of scientific research, but in modern times it patronizes science as a very useful adjunct to metaphysic by showing how specters, as phenomena, symbolize noumena. The assumptions of metaphysic as it patronizes science would be the richest jest of civilization had they not their equal in the ridicule they make in considering realities as base-born, belonging only to the lower world where men live, while metaphysic is supposed to dwell in a region of sublime thought.

We have defined ghosts as fallacies of hallucination conceived as forms. Those who believe in ghosts define them in some other way. Milton may be considered one of the best authorities on ghosts:

> . . . . . . for spirits when they please
> Can either sex assume, or both; so soft
> And uncompounded is their essence pure;
> Not tied or manacled with joint or limb,
> Nor founded on the brittle strength of bones,
> Like cumbrous flesh; but in what shape they choose,
> Dilated or condens'd, bright or obscure,
> Can execute their airy purposes,
> And works of love or enmity fulfill.

Shakspere does not believe in ghosts, but he knows how they are seemingly produced by hypnotism.

Ham.—Why, look you now, how unworthy a thing you make of me. You would play upon me; you would seem to know my stops; you would pluck out the heart of my mystery; you would sound me from my lowest note to the top of my compass; and there is much music, excellent voice, in this little organ, yet cannot you make it speak. 'Sblood! do you think I am easier to be played on than a pipe? Call me what instrument you will, though you can fret me, you cannot play upon me.

Enter Polonius.

God bless you, sir!

Pol. My lord, the queen would speak with you, and presently.

Ham. Do you see yonder cloud, that's almost in shape of a camel?

Pol. By the mass, an 'tis like a camel, indeed.

Ham. Methinks, it is like a weasel.

Pol. It is backed like a weasel.

Ham. Or, like a whale?

Pol. Very like a whale.

Ham. Then, will I come to my mother by and by. They fool me to the top of my bent. I will come by and by.

CHAPTER XXII

FALLACIES OF APPREHENSION

Fallacies have been divided into two grand divisions, which we have called illusions and delusions. It will be remembered that we are reclassifying illusions and delusions, each into five classes. Of the illusions we have already set forth the missensations and the misperceptions, and of the delusions we have set forth the hallucinations and the specters. In considering fallacious apprehensions we discover misapprehensions and phantasms. Let us first set forth the nature of misapprehensions.

We are conscious of pressure when bodies impinge against us, and we are conscious of push when we impinge against other bodies; we are therefore conscious of energy both from an active standpoint and from a passive standpoint. But the energy of which we are conscious is that of molar bodies. We must here recall the fact that knowledge begins in the race and also in the infant with the cognition of molar bodies. To the primitive or naïve apprehension, motion is an effect of a cause, and this cause is considered as something which acts on another and produces motion in self, in order to act on that other, and it may also produce motion in that other. It was long before man cognized that force is itself motion and motion is force. Primitive man formed the habit of considering motion as an effect of force. He was conscious that he could exercise force, and discovered that it could produce molar motion. He knew nothing of molecular motion, or that the force which he exercised was derived from molecular motion, so he considered force and motion as disparate properties; this is the primordial misapprehension.

Erroneous judgments once made may be repeated in perpetuating fallacies, for this constant repetition of fallacious judgments is intuition, and there seems to be something sacred about intuition. A world of metaphysic is built on this foundation, that habitual or intuitive judgments are the primordial endowments of mind. A myth is invented to explain a fallacy, then the myth becomes sacred and the moral nature is enlisted in its defense.

The stars were seen to move along the firmament, or surface of the solid, from east to west, as men move along the surface of the earth at will. But the heavenly bodies move by constantly repeated paths, and so primitive man invents myths to explain these repeated paths. For example, the Utes say that the Sun could once go where he pleased, but when he came near to the people he burned them. Tavots, the Rabbit-god, fought with the Sun and compelled him to travel by an appointed path along the surface of the sky, so that there might be day and night. It is an offense to the religion or moral sentiment of the Ute to question this explanation.

The man is conscious that he can move himself, though he is not conscious that the molecular motion in his body is motion, but he is conscious that it produces the effect of molar motion, and he calls this unknown something force. In what manner this molecular motion of the particles of the body is transmuted into molar motion of the body, is not known except by a few scientific men who see that molecular motion of the particles is transmuted into the molar motion of the body through the metabolism of the muscle, and that this motility or self-activity is controlled by the will which controls the choice or affinity of the molecules of the muscles.

This primordial misapprehension is universal to mankind in tribal society, and universal in explaining all motion. Although not formulated in this manner, it is practically believed that motion, which is simple and well known, is the medium between occult force as one force acts on another. This is a very natural error in the stage of culture to which it pertains.

We speak of the sun, the moon, and the stars as rising and setting, and when the sun rises we conceive it in such terms of speech, but in fact the earth in its daily rotation turns toward the sun. Under favorable circumstances I can see the earth turn toward the sun, down in the front when looking at the sun, and up as my back is turned. I have often experimented in this manner with both the sun and the moon, when I have been traveling on the desert, and I can see their rising and setting as the rotation of the earth. I assure you it is a marvelous revelation. It seems like riding on a Ferris wheel. It is just such revelations as these that a man must experience when he discovers new truths in science. When the fallacy wholly vanishes and the verity appears in all its meaning, it is impossible to conceive a fallacy; but when the fallacy and the verity are both believed, we believe contradictions or antinomies.

Phenomena are expressed in words before they are properly understood; when they come to be known the facts do not properly fit them. I speak of the path of the heavenly orbs extending from east to west, but the fact is that the earth revolves from west to east. The metaphysician takes propositions to express judgments, as they are formed before the phenomena are properly understood by science, to be valid, and then finding that which science ultimately discovers, takes it also to be valid, and discovers in the world a set of contradictions.

Consider a tower a thousand feet high, from which there projects an arm so that a cannonball falling from it will strike the ground outside of the base of the tower. Now let a ball be dropped from this arm, and you say it falls to the ground in a straight line. This is not true; the cannonball and the earth both have the motion of the earth in rotation about its axis; the path of the cannonball, therefore, has two components, one in the direction of rotation and another in the direction of fall. Its path, therefore, is in the direction of fall and rotation. This is not all of the path of the ball: it is moving in revolution with the earth and the moon; it is also moving in revolution with the orbs of the solar system about the sun as the center; it is also moving with the solar system about some point in the galaxy. It falls to the earth, therefore, in a vortical or spiral path, because the earth itself is moving in such a path. For some purposes it is necessary only to consider this movement of the earth as a straight line, because only this component of path must be considered when we consider the change of the ball in relation to objects on the earth, when the real path of the cannonball seems to contradict the considered path, and we have an antinomy.

You say that the book lying on the table is at rest, and you conceive rest as a motionless state. But this is not true; the book which lies on the table has the motion of the earth on its axis, and it has also the motion of the hierarchy of celestial bodies, and it has also the motion of a hierarchy of molecular bodies. Rest, therefore, is only motion parallel to the other bodies of this room, and if you deflect its other motion, so that it is no longer parallel to the other bodies, you produce molar motion. If you still hold that rest is a motionless state, and then apprehend, as you do, that the book is in motion when at rest, you believe contradictions. These contradictions are antinomies. One or other of every antinomy is a fallacy.

If I have set forth the nature of antinomies clearly, I am prepared to set forth the fallacy of Kant's second antinomy. This fallacy consists in holding that there is some force which is not motion, but structure. It is the failure to conceive properly that all bodies are composed of discrete particles which are incorporated by modes of motion, and the failure also to conceive that there is a hierarchy of bodies in which the particle itself is a constituent and that the particle partakes of all the motion of the bodies in which it is incorporated, so that the motion of the particle is vortical. No matter how large or how small the particle may be, it exists in an environment of other particles with which it collides; and by reason of its environment its tendency to a rectilineal path is made vortical, and whenever this vortical path is disturbed by an unwonted collision, it has a tendency to be straightened. Thus the cannonball falling has its path to the earth deflected to one somewhat more in a right line. In order that this statement may more clearly be understood, it requires a further development of the motion of a particle in a hierarchy of bodies. If we can attain to this concept, then the fundamental doctrines of physics are self-evident.

The misapprehensions relating to the forces of molecular bodies linger much longer than those relating to stellar bodies. Only in late years have we learned that heat is a mode of motion, that light is a mode of motion, that electricity is a mode of motion, and a few physicists still believe that gravity is an occult force. Although the law of the persistence of energy or the correlation of forces is established, yet a few apprehend gravity to be an occult force, as attraction and repulsion involving *actio in distans;* yet gravity, when it is understood as a mode of motion, is so simple that all of its laws can be derived by the Euclidean process from the law of the persistence of energy.

There yet remain certain properties or bodies as forces which usually are not conceived as modes of motion. Inertia and rigidity are the two most important. If they are deprived of their occult attributes, all other forces fall into line as modes of motion. Inertia, as defined by Newton, is resistance to deflection of motion, or resistance to acceleration, positive or negative; but when we remember that a body has the internal motion of its parts, and properly conceive that these motions are deflected when the body is accelerated, inertia becomes simple as resistance to deflection. When we conceive that inertia is resistance to deflection, it becomes a proposition, easily comprehended, that rigidity is resistance to the differential deflection of the molecular parts of a body. Every one of its minute parts must be moved if the body is moved, and the regional parts as distinguished from the molecular parts cannot be moved without fracturing the body. Thus we see that rigidity can be explained simply as a mode of motion without resort to occult force.

I am riding in a railway coach. The world moves by. Houses and men are on the wing, landscape and animals are in flight, yet all this motion in the external world is an illusion which I soon learn to correct. I and my railway coach are the moving bodies. Every time I look out of the window I correctly interpret the motion in this manner. My coach stops at a railway station, and the trains near me move. Now, I have formed a habit of interpreting the passing of outside bodies as motion in myself and the coach, and when the trains outside move I infer that I and my coach move, and so strong is this inference that I am impelled to look for some verification before I can decide in which body the molar motion inheres, for the contradictory judgments are both intuited.

It has been demonstrated by science that motion is persistent—cannot be created or annihilated—and the demonstration has been accepted by a great body of scientific men. Antecedently to this demonstration Newton had propounded three laws of motion, one of which is that action and reaction are equal and in opposite directions. In this law the persistence of motion or the indestructibility of energy was implied, but at first its full significance was not understood, perhaps not even by Newton himself.

In the "Principia" his first chapter is a series of definitions, the third of which is as follows:

"The *vis insita*, or innate force of matter, is a power of resisting, by which every body, as much as in it lies, endeavors to persevere in its present state, whether it be of rest or of moving uniformly forward in a right line.

"This force is ever proportional to the body whose force it is, and differs nothing from the inactivity of the mass, but in our manner of conceiving it. A body, from the inactivity of matter, is not without difficulty put out of its state of rest or motion. Upon which account this *vis insita* may, by a most significant name, be called *vis inertiæ*, or force of inactivity. But a body exerts this force only when another force impressed upon it endeavors to change its condition, and the exercise of this force may be considered both as resistance and impulse; it is resistance, in so far as the body for maintaining its present state, withstands the force impressed; it is impulse, in so far as the body, by not easily giving way to the impressed force of another, endeavors to change the state of that other. Resistance is usually ascribed to bodies at rest, and impulse to those in motion; but motion and rest as commonly conceived are only relatively distinguished, nor are those bodies always truly at rest which commonly are taken to be so."

In the last clause it is apparent that Newton himself was conscious of an illusion in the common conception of the term rest, and it is plain from his entire discussion that his term inertia stood for real force, although many scholars since his time have denied this proposition. Had Newton discovered the real nature of what he called *vis inertiæ*, the

"Principia" would have been simplified, as it has been since his time, by definitions given to momentum, energy, force, and power. But even these newer definitions can be revised and the subject presented in a simpler manner.

Vis inertiæ, or inertia, is a component of real force, inherent in every particle of matter as speed of motion, which can be changed in direction only through the agency of collision. The explanation of Newton's third law of motion in this manner changes the ideas of motion as they have hitherto existed in philosophy. Motion as speed is inherent, and not something imposed from without. If, indeed, this be true, then much reasoning in scientific circles must be revised, for it has far-reaching results.

In every mind the term rest seems to imply absence of motion, and thus to have a negative content. This implication still properly remains with the term, and while rest does not mean absence of all motion, it still means absence of molar motion. To the ancients, it meant absence of all motion, and this is the fallacy, but it still means absence of molar motion. My pulse beats as the heart beats and the blood flows. The book on my desk is pulseless; that is, it is devoid of that motion of blood impelled by the heart at every beat; still it has motion, though not pulse motion; so the book which lies on the desk has motion, but not molar motion. As the book is not devoid of motion because it has no pulse, so it is not devoid of motion because it has no molar motion.

Molar motion is the only motion that can be seen directly by the eye without instrumental aid. These molar motions have been so often inferred and verified that the concept is intuitional in every human mind. The concept of stellar motion has also been verified, and the concept is intuitive with some but not with all minds, but the concept of stellar motion has the same validity as the concept of molar motion. The concept of molecular motion, though not intuitional to most people, is just as valid as that of stellar or molar motion.

Concepts of molar, stellar and molecular motion are formed in precisely the same manner by the consolidation of verified judgments. The distinction is not between sense judgments and intuitive judgments, but between verified and unverified judgments, for intuitive judgments may themselves be fallacious.

If I seem to dwell on this point and elaborate the explanation, it is because the illusion of a motionless state must be dispelled before other facts in relation to motion can properly be considered.

An unquestioned fallacy exerts a vital influence on all modes of thought to which it may relate, and engenders a spirit of defense that easily develops into antagonism.

In Spencer's "First Principles," the third chapter is on ultimate scientific ideas. In the seventeenth section he says:

"A body impelled by the hand is clearly perceived to move, and to move in a definite direction: there seems at first sight no possibility of doubting that its motion is real, or that it is towards a given point. Yet it is easy to show that we not only may be, but usually are, quite wrong in both these judgments. Here, for instance, is a ship which, for simplicity's sake, we will suppose to be anchored at the equator with her head to the west. When the captain walks from stem to stern, in what direction does he move? East is the obvious answer—an answer which for the moment may pass without criticism. But now the anchor is heaved, and the vessel sails to the west with a velocity equal to that at which the captain walks. In what direction does he now move when he goes from stem to stern? You cannot say east, for the vessel is carrying him as fast towards the west as he walks to the east; and you cannot say west, for the converse reason. In respect to surrounding space he is stationary; though to all on board the ship he seems to be moving. But now are we quite sure of this conclusion?"

Then he goes on to discuss the motions of molar bodies on the surface of the earth as related to the rotation of the earth on its axis, the revolution of the earth about the sun, and the revolution of the solar system about some point in the heavens lying in the direction of Hercules, but he neglects the molecular motion within the molar body itself. In this discussion he is evidently under misapprehension, which has already been explained and the certitude demonstrated. This certitude is that the acceleration of a body in its proper motion is deflection of its particles. Thus, when a ship is moving in one direction at a certain rate, and the captain is walking from stem to stern at the same rate, his body is deflected by the ship as molar motion in one direction and by motility in the opposite direction; that is, there is a double system of deflection of the particles of his body that compensate one another. The whole subject is thus explained as a double deflection, and all the mystery is solved.

Later in the section Spencer says:

"Another insuperable difficulty presents itself when we contemplate the transfer of Motion. Habit blinds us to the marvelousness of this phenomenon. Familiar with the fact from childhood, we see nothing remarkable in the ability of a moving thing to generate movement in a thing that is stationary. It is, however, impossible to understand it. In what respect does a body after impact differ from itself before impact? What is this added to it which does not sensibly affect any of its properties and yet enables it to traverse space? Here is an object at rest, and here is the same object moving. In the one state it has no tendency to change its place; but in the other it is obliged at each instant to assume a new position. What is it which will for ever go on producing this effect without being exhausted? and how does it dwell in the object? The motion you say has been communicated. But how?—What has been communicated? The striking body has not transferred a thing to the body struck; and it is equally out of the question to say that it has transferred an attribute. What then has it transferred?"

How simple the explanation! Motion as speed cannot be transferred, but motion as path may be deflected.

Then he goes on to demonstrate the absurdities of transferring motion as speed from one body to another, and he finally says:

"Thus neither when considered in connection with Space, nor when considered in connection with Matter, nor when considered in connection with Rest, do we find that Motion is truly cognizable. All efforts to understand its essential nature do but bring us to alternative impossibilities of thought."

In this argument he assumes that the transference of motion is the transfer of speed, but we have demonstrated that the transference of motion is only the transfer of direction by change in the paths of each, which is simple and can be understood by a boy. But the transfer of motion as speed leads to curious and contradictory conclusions, some of which Spencer develops. Here he is reasoning about a fallacy, something which does not exist, and something which is not only unknown, but unknowable, as he affirms. In all of part first of the "First Principles," wherever he discusses scientific subjects, he deals with fallacies and assumes non-existent things borrowed from the history of metaphysical opinion, all involving contradictions, and as no explanation of them can be given, assumes that they are unknowable; still he affirms that they are known as something relative which he explains as something known in a symbolic manner. Now, these fallacies are all represented in literature, and have words by which they are known, but they are symbols of fallacies when improper meanings are given to them, but symbols of certitudes when proper meanings are implied. In all the history of metaphysic I know of no better illustrations of reasoning about fallacies than are here found in this first part, for the propositions are stated with singular clearness; they are never presented in obscure rhetoric, nor are they enforced by an appeal to moral sanctions.

Spencer is right. The doctrine that motion as speed can be transferred from one particle to another is incomprehensible, or, to use his language, is unknowable, or, to use my language, it is absurd. We must not believe incomprehensible, unknowable, or absurd things. Since the days of Euclid, we are accustomed to the doctrine of *reductio ad absurdum* in scientific logic. If we can reduce a proposition to absurdity we reject it.

Spencer goes on in the same chapter to a consideration of force. He says:

"On lifting a chair, the force exerted we regard as equal to that antagonistic force called the weight of the chair; and we cannot think of these as equal without thinking of them as like

FALLACIES OF APPREHENSION 365

in kind; since equality is conceivable only between things that are connatural. The axiom that action and reaction are equal and in opposite directions, commonly exemplified by this very instance of muscular effort versus weight, cannot be mentally realized on any other condition. Yet, contrariwise, it is incredible that the force as existing in the chair really resembles the force as present to our minds. It scarcely needs to point out that the weight of the chair produces in us various feelings according as we support it by a single finger, or the whole hand, or the leg; and hence to argue that as it cannot be like all these sensations there is no reason to believe it like any. It suffices to remark that since the force as known to us is an affection of consciousness, we cannot conceive the force existing in the chair under the same form without endowing the chair with consciousness. So that it is absurd to think of Force as in itself like our sensation of it, and yet necessary so to think of it if we realize it in consciousness at all."

The force in the chair is molecular force; the force in the arm is vital force, partly transmuted into motility, and in the act of lifting the chair molecular force is transmuted into molar force; force in the chair is one mode of force, and in the arm another mode of force; but they are equal, and action and reaction take place, producing effects in opposite directions. The chair moves up, and the man and the earth move down. Of the force in the arm the man is conscious; of the force in the chair he is cognizant, that is, it is learned by combined judgments through inference. But Spencer has never analyzed judgment; he does not distinguish between consciousness and inference, sometimes using consciousness in the sense in which science must use it, but oftener using it in the sense of cognition, and always confounding the two meanings, he rests under the fallacy of the double meaning in consciousness, and reifies it as cognition itself. But the illusion which especially concerns us here inheres in his notion of force. With him force is the ultimate property into which all other properties are resolved, for he seems to resolve kind into force, but of this I am not sure; plainly, he resolves extension into force, by attempting to show that our knowledge of extension is derived from force, not seeing that there can be no knowledge of force without a knowledge of form—that the two are indissoluble properties.

Spencer is supposed to be the philosopher of evolution, and that is his grand theme, but he resolves change into force, not seeing that there can be no change without force, and no force without change. He seems to resolve judgment under the term consciousness, or under the term mind, into force, though his doctrine on this subject is obscure; but with great emphasis and great reiteration, he denies that judgment as mind or consciousness or cognition can be rendered in terms of motion. In this respect he is sound. With him motion is derived from force, not force from motion, and from this force he derives change and persistence; the absolute of change he explains as persistence of force. Then he derives extension from force, and vaguely derives kind from force, and leaves force standing as the substrate of the substrate—the substrate of that which we call matter or substance. Then he argues that extension as a reality must be resolved into void space, and he affirms, without attempting to demonstrate it, that time, as persistence and change, must be resolved into void time, so that with three fallacious entities—void space, void time, and the resolution of all of the attributes of substance into void force—he has three nothings, three voids, three illusions, with which he deals in the first part of his book; and reasoning about these illusions he comes to the conclusion that they are unknowable, but that they are also known in a symbolic manner, and how known in a symbolic manner we have already shown—that it consists in using terms in an illegitimate manner.

It is a dangerous doctrine to claim that we know something because we can talk about it, for we can talk about fallacies and hypotheses as well as about certitudes. Fallacies coined into words or coined into concepts are still fallacies.

In the third chapter of the second part, beginning with the 46th section, Spencer says:

"That sceptical state of mind which the criticisms of Philosophy usually produce, is, in great measure, caused by the misinterpretation of words. A sense of universal illusion ordinarily follows the reading of metaphysics; and is strong in proportion as the argument has appeared conclusive. This sense of universal illusion would probably never have arisen, had the terms used been always rightly construed. Unfortunately, these terms have by association acquired meanings that are quite different from those given to them in philosophical discussions; and the ordinary meanings being unavoidably suggested, there results more or less of that dreamlike idealism which is so incongruous without instinctive convictions. The word phenomenon and its equivalent word appearance, are in great part to blame for this. In ordinary speech, these are uniformly employed in reference to visual perceptions. Habit, almost, if not quite, disables us from thinking of appearance except as something seen; and though phenomenon has a more generalized meaning, yet we cannot rid it of associations with apearance, which is its verbal equivalent. When, therefore, Philosophy proves that our knowledge of the external world can be but phenomenal—when it concludes that the things of which we are conscious are appearances; it inevitably arouses in us the notion of an illusiveness like that to which our visual perceptions are so liable in comparison with our tactual perceptions. Good pictures show us that the aspects of things may be very nearly simulated by colors on canvas. The looking-glass still more distinctly proves how deceptive is sight when unverified by touch. And the frequent cases in which we misinterpret the impressions made on our eyes, and think we see something which we do not see, further shake our faith in vision. So that the implication of uncertainty has infected the very word appearance. Hence, Philosophy, by giving it an extended meaning, leads us to think of all our senses as deceiving us in the same way that the eyes do; and so makes us feel ourselves floating in a world of phantasms. Had phenomenon and appearance no such misleading associations, little, if any, of this mental confusion would result. Or did we in place of them use the term effect, which is equally applicable to all impressions produced on consciousness through any of the senses, and which carries with it in thought the necessary correlative cause, with which it is equally real, we should be in little danger of falling into the insanities of idealism."

Here the confusion which arises from fallacy, together with the contradictions involved, are fittingly set forth; but our philosopher accepts the fallacies and indorses the contradictions, and finally speculates with the difference in meaning between the terms phenomenon and appearance, and he adopts the philosophy of noumenon and phenomenon, and makes the noumenon to stand for the thing in itself the occult force, which he supposes to be void substance and void motion. While Spencer reasons about nonentities or fallacies in his first part, he sets forth many important principles in the second part, but they are all more or less vitiated by fallacies.

How shall we rid ourselves of these fallacies? There is one simple rule. All contradictory concepts must be examined to discover the judgments that lead to contradictions, when correct reasoning

FALLACIES OF APPREHENSION

will eliminate the incongruous. We may always know that concepts are incongruous or contradictory when they lead to a belief in the unknowable. Belief in the unknowable is pessimism about reason and is an evidence of fallacy. Fallacies can be eradicated only by a thorough examination of the concepts involved. The final fallacy on which the philosophy of the contradictory rests can be corrected only by systematic verification of the elementary judgments of which it is composed, and thus by eliminating the errors.

In the 50th section, Spencer says:

"It is a truism to say that the nature of this undecomposable element of our knowledge is inscrutable. If, to use an algebraic illustration, we represent Matter, Motion, and Force, by the symbols, x, y, and z; then, we may ascertain the values of x and y in terms of z; but the value of z can never be found: z is the unknown quantity which must forever remain unknown; for the obvious reason that there is nothing in which its value can be expressed. It is within the possible reach of our intelligence to go on simplifying the equations of all phenomena, until the complex symbols which formulate them are reduced to certain functions of this ultimate symbol; but when we have done this, we have reached that limit which eternally divides science from nescience."

But his letters stand for fallacies; the certitudes should be represented by A, B, and C, then C should be resolved into B, and B into A, as one of the known concomitants of matter.

Bear with me in the reiteration of a fundamental illustration. A and B are particles that collide because they have incident paths. When they collide action and reaction are instantaneous and equal, and no speed is lost in either, but when we consider the antecedent and the consequent as cause and effect, we consider the angle of incidence and compare it with the angle of reflection, and find them equal. If the angle of incidence is 90 degrees, the angle of reflection is 90 degrees, and the particles return reversely by the paths in which they approached. If the angle of incidence is less than 90 degrees, the angle of reflection is less than 90 degrees. If the angle of incidence is one degree, the angle of deflection is but one degree. In all of these cases the force remains equal, and in all of these cases the effect remains equal to the cause, but the force cannot be said to be equal to the cause or to the effect, for the cause is angle of incidence, and the effect is angle of reflection. This simple explanation of the difference between causation and force is a complete refutation of all of Spencer's philosophy of the unknowable. It is also a complete refutation of the doctrine of the dissipation of motion, which he accepts and uses as fundamental to the explanation of evolution.

This is an illusion which we must not neglect. When it is held that motion as speed can leap from one body to another, the doctrine of the dissipation of motion is invented. When the heated iron cools, it is supposed that the iron yields its motion as speed, and dissipates it into surrounding objects, and especially into the ether; it was not seen that the thermal motion in the body is transmuted into another mode of molecular motion still within the body, as exhibited in strength and rigidity. From this fallacy logical consequences are derived when it is held that the sun is dissipating its motion because it is a cooling body. For does not the motion of the sun as heat come through the ether to the earth, and to all other external bodies? Yes, but not as motion, but as cause. Path of motion, not speed of motion, is communicated. The different modes of heat and of light in the ether are not different modes of speed, but different modes of trajectory. Whether the sun can continue to shine is not a question of the dissipation of motion as speed, but a question of the transmutation of one form of motion, called heat, into another form of molecular motion in the body itself. If the conditions for transforming heat into another mode of motion are not favorable to this transmutation, then the sun may still continue to shine and make the planets glad.

Let me suggest, merely as an hypothesis, some reasons for believing that the sun will not go out. On the earth we discover four partially differentiated bodies: air, water, rocks, and the great central body. Geologists have established the theory that this great central body is in a trans-fluid condition, due to pressure, and that thus its heat cannot be transmuted into structural motion. Now the sun is a much larger body than the earth, and for this reason the materials in its outer crust have high specific gravity, and by reason of this higher specific gravity the solid crust must always be thinner, and perhaps this thinner crust cannot be supported against the stresses and strains produced by the stellar motion of the sun, and the stresses and strains developed in the crust itself and coming from the molten nucleus. It may be that the sun's spots, changeable as they are, give evidence of the breaking down, remelting, and reforming of this thin and variable crust.

I do not present this exposition as anything more than an hypothesis, but perhaps it may be considered worthy of an examination by those better equipped for the investigation. If we are to accept the persistence of energy, we must accept the persistence of motion; if we are to accept the persistence of motion, we are compelled to accept the persistence of motion as speed in every particle. Much scientific speculation needs revision.

We must now turn our attention to the fallacies of apprehension, which are derived from hallucinations, and which first become specters, and then in the stage of apprehension become phantasms. By contemplating hallucinations as phantasms, another stage in the development of delusion is produced. When we consider specters in action we consider phantasms.

When we dream we often go abroad, and the specters of our dreams are engaged in activities. It is from this phenomenon that the primitive mind reaches the conclusion that our ghosts may leave the body. Primitive men realize in others, and believe of themselves, that the body remains quiescent in sleep, and to account for the actions of the specters of the dream they conclude that the ghost can leave the body. When this false judgment becomes habitual—i. e., that the property of conception or judgment can depart from the body and sustain an independent existence, without number, space, motion, and time, or in reciprocal terms, without kind, form, force, and causation—then the specters of dreams may have a separate existence away from the body, as shades, subtle forms, or occult personages.

Among tribal men these occult personages usually leave the body by the portal of the nostrils, and return to it by the same gateway. There is a vast amount of lore concerning ghosts and the circumstances under which they leave the body. Stories of ghosts that leave when the body sleeps; stories of ghosts that leave when the person is absorbed in deep contemplation, and the ghost snatches the opportunity to make a journey by itself; stories when ghosts leave the body for the purpose of gaining information in distant parts; stories of ghosts that are sent on journeys by hypnotic suggestion; stories of ghosts that have wended their way to a distant land on wings of magic, at the will of the intoxicated shaman; and stories of ghosts that have permanently left the body and thus have produced insanity, are abundant in the folk-lore of superstitious people. In the same manner the ghosts of others may come to us in our dreams, and be their cause. They may come to us in states of ecstasy, and make us perform many wonderful deeds; they may come to us in hypnotism and become foreign tenants of the body to do their own sweet will; they may come to us in states of intoxication and perform antics in our bodies and revel in delight, for in insanity they take more permanent possession of the body, and our lives will be controlled by foreign residents. It is thus that the actions of men are attributed to ghosts—perhaps wise actions when they go out and return to us with information from the external world; perhaps foolish actions when they take possession of us while our ghosts are away. It is in this manner that many of the mysteries of existence are explained.

CHAPTER XXIII

FALLACIES OF REFLECTION

Fallacies of reflection are fallacies of time and cause, and they may be classed as misreflections and myths. The misreflections are a fourth group of illusions and the myths a fourth group of delusions.

Fallacies concerning time are analogous to those concerning space. Time is persistence and change. It is not blank time, it is a time of something that exists, not the time of something that does not exist. It is the time in which all existence persists and in which it changes. The seed is developed on the apple-tree. Its time is the period of its existence as a germ, but the germ itself was developed by the incorporation of molecules. The molecules existing as particles in the air were transformed into the seed, but the molecules persisted before the seed was formed. The persistence is eternal in the atom so far as we know, but it is changeable from its state in the air or the water into its state in the seed, so its persistence is partly taken up while in the seed state. The seed is planted and becomes a tree by addition of other particles from the air and the water, and the eternal persistence of all the particles is occupied for a period in the state of the tree. Now, the existence of the molecules in the air and the water, and their existence in the seed, and their existence in the tree, and finally their existence as water and air, when the tree is reduced to another state by decay, is a permanent existence, while the temporary existence is in the seed and the tree.

Before man knew that the seed was a continued existence of particles, and that the tree was a continued existence of particles, it was supposed that the time of these existences was limited, and that there was a blank time. Out of this nothing, something was created, and these creations were in continual change, which were called fluxes or becomings. The real nature of persistence not being understood there was assumed to be a persistence which was blank, and the blank was called time. But persistence, not being known, though called time, was held to be the thing-in-itself, which indeed it was in part, and it was called noumenon. When the noumenon was discovered, the idea of blank time was still retained and it was still noumenon, while the real persistence was called a phenomenon. Now it is apparent that this blank time is a fallacy. It was thus, as in this case, that all unknown things, when they came to be known, were transferred to the things which were called phenomena; and the blank things were still called noumena. Thus noumenon was a word originally valid, an x in logical computation, whose value was to be determined; but ultimately it came to mean a something which could not be determined—not only an unknown but an unknowable thing, and a knowable thing was held to be only appearance and was called phenomenon.

My horse is stolen, by whom I know not, and I say there is a thief, but as I do not know this thief I call him a noumenon. But the detectives capture him and he is sent to prison; now the thief becomes a phenomenon, for he is apparent—he may be seen in the jail. Now, suppose that I had talked about this noumenon, when he was unknown, in a conglomeration of attributes—as an uncanny man, as a vicious man seeking another that he may devour him, as a man of seven heads and ten horns; but now I find him only a poor misguided man with the vice of cleptomania or the greed for possession which made him a criminal, but without multiple heads or multiple horns. Having discovered my fallacy in this case I still retain the notion of existence of such a thing as I had imagined, and I continue to believe in it and still call it a noumenon. In the same manner every noumenon of metaphysics can be traced back to the original fallacy entertained by mankind and still supposed to exist as a reality in the universe. When all of these illusions are considered we have the world of occult noumena— the theater of idealism.

Kant explained his occult space, not as a property of physical nature, but as a form of the mind, whatever that may be. In the same manner his occult time was not an existence in physical nature, but also was a form of the mind. He had not the insight to discover that such forms are fallacies, like the dome of the sky in the mind of an ignorant man; still, he had the logical integrity to see that such space and time are incongruous with a space of extension and position and a time of persistence and change, and he boldly followed his logic in formulating a set of antinomies, or contradictions, both of which he seems to have believed as valid.

Kant himself was accustomed to speak of ideas as forms; that is, to speak of one abstract concomitant in terms of another abstract concomitant. For science this habit is fatal. Tropes are good as poetry, but vicious as terms in propositions of logic. Systems of cosmology originate in this manner. In tribal society the earth is made polar from east to west. About this Occidental and Oriental pole a system of worlds is projected—a world of the east, a world of the west, and, at right angles to these, a world of the north and a world of the south, a world of the zenith, and a world of the nadir, with a midworld which is a plane with sides and corners. All the lower tribes of mankind believe in such a world, and there are expressions used in civilized society which are survivals from this stage of belief. To primitive man these worlds are the realities of his cosmology, and he uses these supposed realities as nuclei for many concepts. For example, he formulates social laws as the laws of the east, the laws of the west, the laws of the north, the laws of the south, the laws of the zenith, and the laws of the nadir. Crosses, swastikas, and formulated statements are alike made to conform to this scheme. In somewhat later culture, when a somewhat clearer concept of the midworld exists, and the east, west, north, and south have been explored, but the zenith and the nadir are yet unknown, there still remains a midworld, a heaven above and a hell beneath. Laws and principles are formulated as heavenly or hellish. The transformation of seven worlds into three constitutes one of the most interesting chapters in the history of human opinion. In the seven-world scheme, method of statement becomes a method of philosophy. This fact has abundant illustration. It is the primal vice of classification which was set forth in the chapter on classification.

By a curious mode of expression often, perhaps universally, found in savage society, time is considered to be four-cornered because we measure time in terms of space. We say the sun rises in the east and sets in the west, and that at midday it is in the zenith and at midnight it is supposed to be in the nadir. Some savages will tell you that time is four-cornered, others will tell you that time is round, but that there are four cardinal points of time. Four-cornered time is a firmly established notion among savage and barbaric tribes. Thus time is formulated as if it were space. Many modern physicists mythologize in this manner about motion, being unable to distinguish motion as an abstract property, because motion is formulated in terms of space and force in terms of parallelograms.

Thus a scheme of expression becomes a scheme of reality. When a three-world scheme is substituted for the seven-world scheme, the four worlds are transformed into four substances, as earth, air, fire, and water. Hence the cardinal points of compass become the cardinal substances. The habit of relegating all animals, all plants, all properties, and all qualities to the seven worlds, is continued under the new scheme by making a something like a classification between properties and qualities, and transmuting the properties and qualities to substances or attributes of substances and qualities, to world beings and attributes of world beings. Properties are grouped in fours because there are four horizontal corners of the world, and qualities are grouped in fours because there are four vertical corners of the world as evidenced by time. Thus a scheme of expression becomes a scheme of philosophy. Wet and dry, cold and hot, constitute a scheme of cardinal properties; earth, air, fire, and water, a scheme of cardinal substances; justice, prudence, temperance, and fortitude, a scheme of cardinal virtues.

It is an error of this nature into which Kant fell when he considered space and time as forms of thought. The habit of expressing thought in terms of form led him to the conclusion that space and time, as disparate properties, are identical with thought as a succession of judgments, instead of being concomitant with thought. But more than this, it was the void form and the void space which Kant supposed to be forms which we are compelled to use as *a priori* elements of reason when we consider form and state.

Fallacies of cause occur in every hour of waking life. We attribute effects to wrong causes. We are especially liable to this from the fact that both cause and effect are conditions, and causation is a change of condition from an antecedent to a consequent. The conditions of every causation are multifarious as we look at them in a regressus of causes or a progressus of effects, and as the mind of the individual can make but one judgment at a time, it may be that the one of the causes or effects which is considered, is in fact a trivial element in the causation, for in all our language we are accustomed to speak of one of the causes as the special cause, for it must be the special one in consideration.

Forces are often processes in which a multitude of unseen objects produce a seen effect, as when many molecules of air strike upon a tree which bends before the blast, or when many raindrops, that can scarcely be seen where they fall and are wholly unseen by the man who beholds the river, create a flood that endeluges a valley.

Some instances of this kind produce fallacies that are widely entertained; they are misreflections that substitute the effect for the cause. One illustration of this group of fallacies must suffice for us here. Some years ago there was published an interesting and well written book, the theme of which was the origin of deserts, giving a pessimistic view of the world, in which it was represented that desert conditions are increasing, and that wide regions of country have already been laid waste as deserts, because mankind interferes with the operations of nature by destroying the forests, and that if forests were restored rainfall would be increased. In this manner effect was taken for cause.

The most subtle fallacy about causation consists in mistaking it for another property, either as force on the one hand or as thought on the other. Force, cause, and conception—or motion, space, and judgment—are disparate properties but concomitant in every particle and body of the universe. This has been the burthen of our theme from the chapter on essentials, in which it was affirmed, to the present one, and all our demonstrations have had this end in view.

He who cannot clearly distinguish between abstract and concrete, or between body and property, is certain to fall into mysticism. Mill and Spencer in the late years, like Aristotle in ancient time, confounded causation with force or energy, while Kant and all the school of metaphysicians confound both cause and force with thought.

Evolution is a succession of changes which are in time and require time for their accomplishment.

The ancients believed and the tribes believe that kinds, forms, and forces come out of nothing and return to nothing. This is the primal fallacy of causation. Modern science has demonstrated that kinds, forms, and forces come from something else and vanish into something else. It is only today that this is universally accepted by scientific men, while even at the present time millions of those who inhabit the earth still believe in creation from nothing. We shall not attempt to recount the multitude of fallacies which have existed and which still linger in scientific circles. We have already set forth the one most important to our argument, that is, that motion is created by or comes out of some occult force which is not itself motion, and the other form in which motion is supposed to leap or creep, or in some other manner to be transferred from one body to another. An acrobatic motion is the last ghost of force.

We now come to the second part of our chapter, the discussion of myths. Mythology is the history of ghosts. Ghosts are specters, and we have seen what strange acts they commit as phantasms, when they leave the body and travel abroad in the world and return again to the body, or when from abroad they enter the body to take possession of it in the absence of its owner. In savage society authority is wielded by the oldest man, who thus by superior age, natural or conventional, becomes the chief. In the same manner the dwellers in ghostland are ruled by tribes; the progenitor, prototype, or elder animal of the tribe is its chief.

Now we are to consider what it is that ghosts have done—how they have acted in the theater of the universe. Strange to say, we find it well recorded, for ghosts have had more complete recognition than men in all ancient history. Ghosts, as a race, have passed through interesting stages of history. All changes are in time and require time to become discrete quantities of change that may be recognized.

Hence it is that in the evolution of ghosts we have to consider their transmutation from one to another as it appears when we consider them separated by many centuries of time. We are unable to find the distinction in the race of ghosts, if we consider them yesterday and again today, or last year and again this year, or even last century and again this century; but when we consider them as they appear in the stages of culture which are designated as savagery, barbarism, monarchy, and democracy, we find discrete degrees of evolution.

It is only in such considerations that planes of demarcation can be discovered. I shall therefore consider ghosts as they appear in savagery, barbarism, monarchy, and democracy, or to use more common terms, civilization and enlightenment.

In savagery the ghosts are zoömorphic. All lower animals, stones, bodies of water, the sun, the moon, and all the stars are supposed to be animals. The universe is a universe of animals living in the seven regions. All of these animals have ghosts which can leave their bodies and journey through the world, and at will inhabit other bodies, when they find them vacated by their proper ghosts. It is thus that the primitive mythology is a theory of animal ghosts. What these ghosts can do in their proper bodies is easily seen, though it is very wonderful; but what they do when they leave their proper bodies is mysterious or occult.

To the savage, lower animals seem to have attributes and to perform deeds that are more wonderful than those of human beings. The serpent is swift without legs, the bird can revel where man cannot go—through void space with wings. The fish can inhabit the water and run with fins; no human being can do this. The spider can spin a thread and travel on it; all that he has to do is to spin the thread from his own body and travel wherever he wills as it is unwound. The rivers are born of rain and roll into the sea which never increases. The winds are created by the breath of beasts or rise from under the wings of birds from nothing. The stars can fly like birds and shine like fire. So the savage man considers the molar bodies of the world, which are all animals like himself, to have many magical or occult attributes which are very wonderful. But the wonderful things which they do are not attributed to their bodies, but to their ghosts. The body of a man lies inert when he sleeps, but his ghost cannot sleep, it travels about the world when his body is at rest. The bodies of the rocks are inert, but when they sleep at night their ghosts shine in the heaven as the aurora borealis. If you strike one rock with another you can see its ghost as a spark of fire. When the clouds gather they are the ghosts of water; when angry they shine with lightning light, and when pleased the clouds shine as rainbows. These illustrations will serve to show how thoroughly, in the notion of the savage, ghosts and bodies are differentiated.

The universe being considered as bodies and ghosts, and the bodies being considered as inert and the ghosts as active principles, we have the fundamental theory of savage reasoning. We can do nothing except as it is done by our ghosts. We cannot cause anything to be done by others except by controlling their ghosts. Words cause other human beings to do things, and their words cause us to act. The words of the mother cause action in the babe; the voice of the babe causes the mother to act. The voice of the bird brings its mate to its side, or the voice of its mate takes the bird to its side. The primeval concept of causation is the notion that words produce effects, and that effects are caused by words. The bird flies to its mate; the flying of the bird is considered the action of the bird, but when it flies in response to the call of its mate the call seems to be the cause of its flight. It is the special cause; primitive man has no insight into the many causes that are involved. It is from this primeval concept of cause as some special condition, that is developed through the ages, when in a higher civilization we consider the special cause as if it was the total cause. Now mythology, having ghosts as actors, secures their action by causes, and explains the phenomena of the universe as the activities of ghosts acting through body by verbal causation. In savagery words are the ordinary observable causes and constitute the primal cause.

We do not know the languages of the other animals, we can speak to them only through signs or symbols. Great is that man who can talk to ghosts. The symbol which he uses is called a mystery. In the Ute language it is *pokunt;* in the Siouan language it is *wakanda;* in the Algonquian it is *manito.* All tribal languages have a word which signifies the mystery, which can be used as a symbol to cause the action of ghosts. The concept is born in savagery of a mysterious cause which has power over ghosts, which again have powers over bodies, and so the universe is a realm of bodies, ghosts, and mysteries, or unknown tongues.

The mystery, called by various names among American tribes, is usually translated "medicine," for the early missionaries found the people appealing to the mystery to heal disease, for diseases are supposed to be ghosts of animals. As the mystery is something which must act as a word, it must be something which will suggest to the ghost that which is wanted. Hence there arises the doctrine of signatures, which means among the tribesmen much more than the signatures of medicines, by which we are to learn what medicines are good for diseases—it primarily means what signatures can be made to convey our commands to ghosts. As ghosts are all animals in savagery, how can we talk to the ghosts of animals? This leads in savagery to the symbols which constitute the paraphernalia of altars. In savagery every object on the altar is a sign to ghosts of what men wish when they perform ceremonies. They pray to the ghosts for rain, and to make sure that the ghosts will understand what they mean, they refer them to cloud symbols. When they pray for corn they place ears of corn upon the altar. When they pray that the corn shall ripen and become hard they place crystals of quartz upon the altar. In various ways signatures are used by the priests in invoking the aid of ghosts. Those persons who have power over ghosts are medicine men or priests, and attain great influence and sometimes are greatly feared. If they use their power for evil, they are wizards and are killed. If they use their power for good, they may be made chiefs.

Primarily the name given to a body designates some property of that body. After a time the name itself becomes the property of the body, and finally the name becomes a mythical body. These stages in the development of words can be discovered in many of the languages of America, doubtless in them all; it is the transmutation which Max Müller calls a disease of language.

In the second stage of culture, called barbarism, animals have been domesticated and thus by more intimate acquaintance with animals the lower animals are dethroned and human animals are exalted. All animals and other molar beings which are supposed to be bodies movable by human beings, are still held to have ghosts, but their rulers are ghosts of human beings and the great phenomena of nature are personified as human beings; the sun, moon, and stars are exalted in this manner; the seas, the rivers and the mountains are likewise personified. All the most important phenomena of the universe as they are known to man are personified. The rising and the setting of the sun, or the dawning and the gloaming, are personified as well as the sun itself. The rainbow also is personified. Fire is personified. The ghosts are no less multitudinous, but some are exalted above the others, and those promoted in this manner are deities of higher rank. To these deities are attributed the important events of the worlds. But there are many minor ghosts; the worlds are full of them, born of the ages.

Now, in barbarism ghosts are still the actors in the worlds and they are caused to act by signs, and tribesmen still continue to ransack the earth for signatures. Men still hold in love or fear those who have the lore of ghost science. The chiefs or head men or ancestors of the ghosts are greatly revered as gods, and common folk ghosts take part in the affairs of the worlds, and mythology is the history of their doings. These folk-talks elaborately portray the life of men and ghosts and the potency of signs.

The ceremonies of supplication which still continue from savagery, are believed to have still more potency by reason of the sacrifices that have become more and more important in the estimation of the people as time has advanced. In savagery the ceremonials are chiefly terpsichorean: music and dancing were the agencies by which the attention of the ghosts was obtained. While in savagery the pouring of oblations and the presentation of the corn were signs of what was desired, and all the paraphernalia of the altar that represented the thing for which men prayed were merely significant of the things men wanted, in this higher stage men have come to believe that the good things which men want are the good things which the ghosts want, only they want the ghosts of the good things, not their bodies. So the altar of signatures gradually becomes the altar of sacrifice. Hecatombs of beeves, bottles of wine, all the first fruits of the harvest, everything the ghost desires, even human beings, may be sacrificed upon the altar. If after this statement my reader will consult the Odyssey he will there find the most vivid portrayal of barbaric philosophy that has been preserved to us from antiquity.

In despotism, or the third stage of social organization, ghosts are still more exalted, in that the psychic characteristics of men are personified. Certain of the gods of barbarism gradually become representatives of certain psychic characteristics, and we have the stage of psychotheism, and there is a god of War, a god of Love, a god of Hate, a god of Commerce, and many other major deities; but there is a second class of deities representing what are supposed to be secondary attributes of human and divine ghosts. It is in this stage that we observe the transmutation of words into gods. The concepts of which words as signs are personified, as Max Müller has abundantly shown. "In the beginning was the Word, and the Word was God." A development of cosmology which begins late in barbarism is more thoroughly carried out. The cardinal worlds are wholly thrown out of mythology and the midworld has a world above or a heaven, and a world below or a hell. The midworld becomes the sole theater for the development of ghosts by birth. These ghosts, born in the midworld of human beings, are the ghosts of the external world which ofttimes visit the earth. The three worlds of the stage of despotism constitute the fundamental schematism of the philosophy of the period. Institutions are of heaven or of hell, opinions are of heaven or of hell, and in all philosophy the schematism prevails. But in this midworld the ghosts of heaven and the ghosts of hell take part with the embodied ghosts of men in all of the affairs of the world. Everywhere there is a ruler, a despot—a commander-in-chief of the hosts of heaven and the hosts of hell; while on earth in the midworld it becomes the ambition of every despot or emperor to become the sole ruler. The ghosts born on earth depart to the upper or the lower regions, where they are forever separated by an impassable barrier, and life on earth is but a probation in which ghosts are selected for the other world; hence the chief purpose of life in the body is attained by securing a happy life in ghostland.

During all this stage in mythology the ghost-gods are affected by psychological considerations. The supreme being in every religion of despotism is especially influenced by the opinions of his followers. Their opinions of the supreme being must be sound, and worship is by faith in spirit and in truth. Thus worship is fiducial. The supreme being is supposed to take delight in the opinions of his followers and in the expression of those opinions as formulated in creed and especially as formulated in ceremony. This mythical stage gives rise to a vast body of folk-lore, which is distinguished from mythology proper by the belief in a ghostly, supreme being. The midworld is still the theater of ghosts who come from the world above and the world below and sometimes dwell for a time in this world and take part in the affairs of men. These ghosts are especially amenable to deeds of necromancy, the more refined form in which the doctrine of signatures is held. If my reader will carefully study Tasso in "Jerusalem Delivered," he will there find recorded one of the best accounts extant of the necromancy of the despotic age. The publications of the various folk-lore societies of the world are rapidly putting these superstitions on record.

I shall refrain from discussing the fourth stage of ghost-lore. In very modern times it has assumed a special phase which is called spiritism, and attendant upon the theory of spiritism there is developed a claim for a scientific explanation of spiritism in the theory of telepathy, which I cannot wholly overlook and do not wish to ignore, but on that phase which is specially represented in religion I purposely remain silent, lest I should antagonize, with my own opinions, the views of others about religion, and thus enter a field of theological disputation. Yet without expressing personal opinions about the evolution of religion, which I have elsewhere done, I shall content myself with only one paragraph upon the subject.

From the doctrine of signatures there has grown the science of modern surgery and medicine. I do not despise the early efforts of mankind to relieve their sufferings, even though they entertained many fallacies; but I rejoice in the outcome of this effort as it is exhibited in modern medicine. Astrology was necromancy at one time, but has become astronomy in modern times, and I look upon the efforts which were made in former times by astrologists as the planting of the germs of the celestial science. So I look upon mythology with no feelings of hatred, for it seems to me to have made great strides in the science of religion or ethics, out of which shall come a purified science of God, Immortality, and Freedom.

CHAPTER XXIV

FALLACIES OF IDEATION

Fallacies of ideation constitute a fifth grade, which are illusions and delusions. In the order heretofore followed, we shall first speak of illusions, and then of delusions.

The Schoolmen speculated much on the nature of kinds, and finally reached the conclusion that that which makes a thing a kind is its essence, i. e., that which is essential to its existence as a kind, like others of its kind, but different from other kinds. All of this is quite true, but it adds nothing to knowledge, except that it might be given as a definition of a word. For a long time definitions were considered very good explanations.

When chemistry was yet alchemy, attempts often were made to discover the essence of things, and, in particular, it was a favorite method to extract kinds, and these extracts were called essences. So the kind or essence of a thing discovered in this manner was supposed to be its essential quality, as this term was then used. We have a record of this superstition, as it existed in the days of alchemy, in the extracts of the apothecary shop, which are often called essences. Rose-water was the essential extract of the rose, violet-water of the violet, and men were pleased with the idea that they could make of that which constitutes a thing or kind a decoction for a lady's dressing table.

Fallacious theories of kind have high antiquity. It has already been set forth that a classification of properties and qualities is made in tribal society by a schematization of worlds. Not only were molar bodies, which were supposed to be animate, classified in this manner into seven categories, but all attributes of bodies were in like manner classified. We have already seen how space properties gave rise to a cosmology of seven regions. We have also seen how motion was explained as the self-activity of molar bodies, and that the heavenly bodies, which were supposed to be molar bodies, are in motion by appointed paths established by conflict in war, and given spacial or world directions, and that force was considered as will and the cause of motion.

We also have seen the manner in which time was considered as an attribute of space. We likewise have seen the development of the seven worlds into three, as the midworld, the zenith, and the nadir worlds.

Here we must pause for a time to explain something more of the nature of this transmutation. The change developed in later barbarism and earlier civilization was wrought by the increase of geographical knowledge. During this period there gradually was developed a notion of the land, or midworld, as a plane from which mountains and hills stand in relief, surrounded by the sea. Thales gives us such an account, as do many others. All the mythology of the time assumes the existence of the midworld as an island surrounded by an ocean. During the same epoch in human culture the unseen atmosphere was discovered. As the cardinal worlds were gradually abandoned, these properties and qualities of bodies that had previously been classified in the world scheme, came to be classified by a very natural change, as attributes of molar bodies. The schematization still remained fourfold, but molar bodies were considered as kinds, composed of four occult substances—earth, air, fire, and water. Thus, the four regions were transmuted into the four substances. Greek philosophy began with this theory, and there is abundant evidence that other races entertained the same doctrine.

Thus, the most ancient philosophy of civilization started with a theory of three worlds and four substances. We must now rapidly trace its development through five stages, during a period of more than twenty centuries, as it is revealed to us in the history of metaphysic as distinguished from science.

We must consider a little further the misunderstandings of ideation. Every man for himself verifies the current judgments which he makes in relation to practical affairs. If our judgments were not verified until after they are acted on, the race would be overwhelmed by disaster. We have already seen that erroneous judgments vie in multiplicity with valid judgments. If a man should act on erroneous judgments, they would lead him into such mistakes that, almost every hour in the day, he would perform some act causing irreparable mischief. The food which he selects must be properly chosen, but the many things which he might select for food, which are injurious, or even deadly, outnumber the articles which should constitute his proper food. The snares, the pitfalls, the precipices, the floods, which beset his path, are so many that his way must carefully be chosen. The forces which are encountered, as men, beasts, and natural powers, are so many that he must constantly avoid antagonisms. Life is a perpetual exercise of choice. Judgments that are made must be verified in practical affairs, lest the race should become extinct. So, in the making of our judgments, we form a habit of verifying them before we proceed to act.

The immediate judgments of practical life *must* be verified, but the judgments which we make about future events may be postponed, and, practically, they are postponed in tribal society. But men come at last to seek for the verification of judgments which are more and more remotely practical, for they also are found to involve ultimate welfare. Then science is born, for science is knowledge, or verified judgments. When science is born, civilization begins. If judgments are incongruous, somewhere there must be error. This is the method of discovering error, which is habitual or intuitive in mankind, developed from infancy in the individual, and developed in the race from generation to generation, through the whole period of animate existence. It is the most profound intuition of the human mind.

With civilization there springs up a philosophy of monism, which is a philosophy of the error involved in judgments that are incongruous. The key to the meaning of that which we call ancient philosophy is found in the attempt to discover a unifying principle.

Through the centuries this has been the quest of wise men. The seemingly multitudinous properties and qualities of body must be reduced to some unifying principle. As the individual first guesses and then verifies, as already set forth in the chapters on intellection, so the race, at one time and another, guesses, chooses, selects some one property to which all other properties may be reduced as the unifying principle.

This quest started at the beginning of civilization, when four substances, earth, air, fire, and water, were held to be the elements of which all bodies are composed. Civilization inherited a controversy from barbarism about these substances. The substances themselves were derived from the cardinal points, and the brotherhoods of the tribe were organized to represent these cardinal points. Each brotherhood claimed for itself an origin in the cardinal point from which it was named, and hence there was a perennial controversy between the brotherhoods as to the most noble or honorable of these origins. Now, in tribal society, the most noble or honorable is the eldest, for that is the method of expressing nobler; elder and nobler are synonymous, for the elder has dominion in tribal society. In the beginning of Greek philosophy, the ἀρχή, the first, held dominion, and was hence the most honorable. The controversies about the most honorable of the points of the compass, the one which should hold dominion, the one which was the first, held over into the stage when the cardinal points were considered as substances, and hence the Greeks inherited the controversy about the first, or ἀρχή, of the elements. Now, another method of expressing this idea is that the first is the grandfather, so that it is customary in tribal society to speak of the chief as the grandfather. The totemic head of the tribe is often called the grandfather, as is also the totemic head who is the first of the tribe, or the one from which all the other members are derived. These doctrines are thoroughly ingrained in the habits of thought and the methods of expression current in tribal society, and inherited by national society. Hence, we find, in the study of Greek philosophy, which primarily is cosmology, the first, or ἀρχή, of the elements still to be the subject of dispute, and the first is taken as the one from which all others are derived, and hence to have dominion, and so the most honorable.

At last, there arose a philosopher who cleared his imagination of the fallacies of kinds, as earth, air, fire, and water, and made the bold hypothesis that all things are ultimately founded, not on kind, but on its reciprocal, number, for Pythagoras was a mathematician. Then began the theories of reified abstractions, the theories by which the properties of bodies are unified as a foundation for monistic philosophy.

There is so much of truth in the philosophy of Pythagoras, that when we consider the universe as composed of properties that can be measured, and that by measure all properties are reduced to number, then all properties can be considered as number. Counting on the human abacus had now been developed into the science of reasoning by conventional numbers, and, having discovered that sound is a numerical relation of vibrations of the air, and carrying his magical philosophy into all his ways and thoughts, but not clearly understanding the nature of measure itself, that it is the rendering of one property into the terms of another, until all of the properties are reduced to number, he conceived the doctrine that the universe is a world of numbers; and so it is, but it is much more than a world of numbers, as we have abundantly seen.

Pythagoras is said to have taught this doctrine. This is not known from records left by himself, but mainly from records which come from his immediate successors. The literature of the Pythagorean philosophy is meager, yet, from the little that remains, it seems to have been a theory of the origin of all other properties from number.

In mathematics, the science of verification is space reduced to number; motion is reduced to space, and then to number; and, finally, time is reduced to motion, and motion to space, and space to number; and all of these conventional reductions are accomplished by the device of measure. But, in the doctrine of Pythagoras, number seems to have been held as the substrate of properties. It is the patriarch of the illusions of metaphysical philosophy; its venerable form is gray with the mystical shadows of twenty-five centuries. This may be denominated the fallacy of Pythagoras.

Plato taught that form is the substrate of all properties. This he did with such literary skill that he held the judgment of mankind for many centuries. He not only taught that form is the substrate of physical properties, but also of thought. To him thoughts were forms given off by objects floating in the empyrean and taken into the mind, and his exposition of this doctrine transferred the word *idea* from the realm of space to the realm of mind. A monument to this fallacy still exists in the use of the term *idea* for a notion in every modern language of civilization. This may be denominated the fallacy of Plato.

Aristotle rejected the Pythagorean and Platonic fallacy, but entertained one of his own. He reified energy or force, which is derived from motion, and taught that this energy is the substrate of all properties. Now, while this seems to have been his doctrine, yet it must be remembered that Aristotle was a careless writer, heedless of the niceties of expression, and unconscious of the necessity for using scientific accuracy in terms. It seems possible to refer to Aristotle as an authority for many of the fallacies which have been entertained in metaphysic, and philosophers usually reverence him as the Master. If I were called on to point out the fundamental doctrine of Aristotle, I should cite his theory of energy, so I call this the fallacy of Aristotle. As his exposition of the subject is not very lucid, and as men may honestly controvert any statement made of his doctrine, it seems better to look for another master of this doctrine. In Spencer, we have a philosopher who rivals Plato in literary skill. In Spencer's "First Principles," where he lays the foundation of his philosophy, he sets forth the doctrine in no uncertain terms. Motion is derived from force, extension also is derived from force, and, finally, all of the properties are held to have force as their substrate. If the reader will consult Spencer's "First Principles," part II, chapter III, he will there discover his method of explaining properties. The chapter is entitled, "Space, Time, Matter, Motion, and Force." He not only derives all the properties, but all bodies, from force, and then describes force as something unknown and unknowable; so, in the name of science, he meets the metaphysician on his own ground, and sets forth his doctrines with a deftness and simplicity with which the dealer in mystery cannot vie. Spencer not only entertains the fallacy that properties are derived from an unknown and unknowable force, but he makes force the substrate of all relations, and then affirms that we know only of relations, and that their substrate is the unknowable; but still more, he accepts the Kantian illusions of a void space and a void time.

Then, time was held to be the unifying principle of all properties and bodies. This reification was designated by the term *being*, taking the participle of the asserting word *to be*, but using it in its secondary sense as signifying to exist. My reading does not furnish me with the knowledge necessary to say who first clearly propounded this doctrine, but it was almost universally entertained by scholastic metaphysicians. Let us, then, denominate this the scholastic fallacy.

It appears that Plato and Aristotle have been recorded more generously than other philosophers of Grecian history. The authority which they wielded seems not to have permitted the revival of the Pythagorean fallacy which they successfully dispelled, while the Aristotelian fallacy had no extensive following until modern times, when, under the lead of Spencer, the great modern master, it has been extensively taught.

But, of all these fallacies, that of the reification of time has, perhaps, had the greatest following; it is the philosophy of Ontology. It has one variety which almost equals in importance Ontology itself. This variety of the species is the metaphysic of becoming, or, as it is sometimes called, the metaphysic of essence, which has many phases, the most important of which is that the essence of a thing is that into which it will develop.

Thus, the philosophy of bodies assumed the phase of substance and properties. How long it held the judgment of mankind is shown when we remember that even Newton himself believed light to be corpuscular emanations from bodies. The last vestige of this doctrine remains when it is supposed that motion jumps from one body to another; and this doctrine is accompanied by another which affirms that path is motion itself. This doctrine of essence is the doctrine which Hegel, in the third chapter of his "Phenomenology of Spirit," sets forth as one of the inadequate judgments of men, which is properly understood only when the external world is considered as a form of thought. There is a curious error prevalent in scholastic times, which is the fallacy of substrates. It was involved in the philosophy of Pythagoras, Plato, and Aristotle, when one of the properties was held to be the substrate of all the others. But it had a long history, and assumed many phases, one or two of which must briefly be set forth.

It was the theory that substance, or substrate, or essence, by whatever name it may be denominated, is porous, and that properties emanate from its pores; that substance gives off an inexhaustible supply of properties. Plato thought that properties were given off from the substance of bodies as forms. For a long time it was held by philosophers that force was thus given off from bodies as subtle emanations.

In this stage of speculation, properties were called accidents, and the theory of bodies took this phase. Bodies are composed of substance and accidents; the accidents may come and go, but the substance remains. John Locke put this subject in a nutshell:

"They who first ran into the notion of *accidents*, as a sort of real beings that needed something to inhere in, were forced to find out the word *substance* to support them. Had the poor Indian philosopher (who imagined that the earth also wanted something to bear it up) but thought of this word substance, he needed not to have been at the trouble to find an elephant to support it, and a tórtoise to support his elephant: the word substance would have done it effectually. And he that inquired might have taken it for as good an answer from an Indian philosopher,—that substance, without knowing what it is, is that which supports the earth, as we take it for a sufficient answer, and good doctrine from our European philosophers,—that substance, without knowing what it is, is that which supports accidents. So that of substance, we have no idea of what it is, but only a confused, obscure one of what it does. . . .

"So that if any one will examine himself concerning his notion of pure substance in general, he will find he has no other idea of it at all, but only a supposition of he knows not what *support* of such qualities which are capable of producing simple ideas in us; which qualities are commonly called accidents. If any one should be asked, what is the subject wherein color or weight inheres, he would have nothing to say, but the solid extended parts; and if he were demanded what is it that solidity and extension adhere in, he would not be in a much better case than the Indian before mentioned, who, saying that the world was supported by a great elephant, was asked what the elephant rested on; to which his answer was—a great tortoise: but being again pressed to know what gave support to the broad-backed tortoise, replied—*something, he knew not what*. And thus here, as in all other cases where we use words without having clear and distinct ideas, we talk like children: who, being questioned what such a thing is, which they know not, readily give this satisfactory answer, that it is *something:* which, in truth, signifies no more, when so used, either by children or men, but that they know not what; and that the thing they pretend to know, and talk of, is what they have no distinct idea of at all, and so are perfectly ignorant of it, and in the dark. The idea then we have, to which we give the general name substance, being nothing but the supposed, but unknown, support of those qualities we find existing, which we imagine cannot subsist *sine re substante*, without something to support them, we call that support *substantia;* which, according to the true import of the word, is, in plain English, standing under or upholding."

It is this something, we know not what, of which Locke speaks, that has come to be designated in metaphysic as noumenon, while the accidents of his time have come to be designated as phenomena. By the Greeks, the fish, seen by its ripple in the water, is called a phenomenon; after it is caught and the fish itself is seen, instead of the ripple, it is called a noumenon. In modern metaphysic, both are called phenomena. The multitudinous properties of bodies can all be resolved into the five essentials which we have set forth; these are the noumena, while the multitudinous phenomena are the relations of particles or bodies to one another. Noumena are constant or absolute; phenomena are relative or variable. This leads us to the discussion of the delusions of ideation.

During the stages of opinion which were characterized by a belief in the Pythagorean, Platonic, Aristotelian, and Scholastic fallacies, as they have been described above, science and metaphysic proceeded together, hand in hand, in search of the truth, though the science of reality was clouded with the metaphysic of fallacy. But now science and metaphysic part company. In this new stage, not only does metaphysic reify, substantialize, or hypostasize the essentials or noumena of consciousness, but it adopts the ghost theory, for the psychic property is considered as a ghost which can leave the body and return to it. The completed stage of the ghost theory is idealism. Opposed to idealism or the ghost theory of spirit—mind or consciousness—is the theory which is most commonly called materialism, of which Spencer is the modern champion.

Idealism began with Berkeley, but he formulated it as a system of theology, or an explanation of the origin of the world in the thought of God. Berkeley gave us, in clear and beautiful English, a theory of vision which was the germ of a new psychology developed by Helmholtz into a more scientific form, with greater exactness, as a scientific theory of vision and also of audition. So Helmholtz may be considered as the founder of scientific psychology. But the idealism of Berkeley was taken up by many others, especially by the German school, represented by Kant, Fichte, and Schelling. Kant, who was the founder of this new German school, left the subject in an attitude wholly unsatisfactory to the human mind as a theory of monism. In his great work, "The Critique of Pure Reason," he pronounces sentence on human reason by consigning its conclusions to the limbo of antinomies; in his subsequent work he relegates man back to practical reason, that is, the formation of judgments which must be made in order that we may act, instead of what we may know. I have already mentioned the primal fallacies into which Kant fell; but he did not produce a system of idealism, nor did Fichte nor Schelling. It was left for Hegel to create a system. This he did by creating a logic of contradictories.

Perhaps I have sufficiently set forth the nature of conception, through the forming of judgments of sensation, perception, apprehension, reflection, and ideation, as stages in the process of forming concepts, and that until all of these stages have been passed there is a probability of entertaining fallacies, especially when we do not recognize that cognition is never completed until judgments are verified. The nature of conception or reasoning, as thus set forth, seems to have been understood by Hegel in some vague way. Hence, he properly explained antinomies as the final harmonizing of judgments by the last process in conception as ideation. So far, I believe his work to be sound; surely it possesses this germ of truth. But he did not clearly understand the nature of ideation, for he was an idealist, and reified the property—the psychic property—of bodies; he was a monist of an abstraction, and he believed the external world to be a fallacy—a phantasm, an illusion, a delusion if you will—something which does not exist in itself.

Hegel does not affirm but he always assumes that there is no external world, that is, there is no reality in the four mechanical properties of body, the four essentials—unity, extension, speed and persistence. They exist only as attributes of consciousness. Consciousness, or idea, to use his term, is the substrate from which flows the accidents or mechanical properties, as from its pores, in an inexhaustible supply of fallacies. There is kind, form, force, and causation of conception, but there is no kind, form, force, and causation, except that which is ideal; that is, he everywhere assumes, and practically affirms, that the mechanical properties are the creations of the mind. It is in this sense that he denies the reality of the external world. Kant gives four fundamental antinomies; but with Hegel all reasoning about the external world, or the four properties of bodies, is fallacious, and the only way to cognize reality is first to cognize consciousness in all its developments, and then to cognize the external world as a system of fallacious judgments. Real cognition must be of the "idea" itself. This is the fallacy of Hegel.

Kant resolves the world of thought into antinomies of contradictions, and refers us back to the practical judgments of good and evil, which control our acts; but Hegel develops a system in which he refers all of our judgments of an external world to fallacies. The only realities or cognitions are those about "idea," as he calls it, or those about consciousness, and its development into the faculties of the intellect, as herein set forth. According to Hegel, the only noumenon is the idea. Mechanical properties of bodies are but phenomena. There are no stars, and we only fallaciously think there are stars. There is no atmosphere, no sea, no formations, no rocks, no nucleus; we only fallaciously think that they exist. There are no plants; we only fallaciously think there are plants. There are no animals; we only fallaciously think there are animals. But there are minds, which, by some occult process, exist not in time or space, and in this occult sense are internal, whatever that may be. The furniture of the world, which we suppose to be external, does not exist, except as fallacy, or, as Hegel calls it, phenomenon. Antinomies arise, when we consider them as realities. But antinomies disappear, if we consider them as ideas.

Having made this discovery, he announces it in the "Phenomenology," and shows us how he reaches this conclusion in a marvelous collection of sentences, paragraphs, and chapters, which, to the scientific mind, at first seem wholly incomprehensible, for the argument is hieratic. It cannot be understood except by those who are initiated into the mysteries of its symbolic language. Though tempted to analyze it, I must not, for it would require a treatise in itself equal to that needed for the unraveling of the cuneiform inscriptions. However, I think that I may pause long enough to show the fundamental principles on which he proceeds: (1) He assumes that mind is the substrate, and hence the unifying principle. (2) He sees as clearly as may be from a study of language, that one property may be spoken of in the terms of another; thus a space may be spoken of in terms of number, as, the distance from the Capitol to the White House may be six thousand feet. We have already seen that measure itself is primarily the reduction of space to number; it is then the reduction of motion and space to number; it is then the reduction of time to motion, and motion to space, and space to number; it is then the reduction of judgment to time, and of time to motion, and of motion to space, and of space to number. These reductions are woven into all the language of daily life, making them tropes, or giving them vicarious uses; but especially do we use terms of the mechanical properties when we speak of the properties of consciousness. The very same words that we use to speak of the properties of consciousness, we more often use when speaking of other properties. It is that which we have set forth as the vicarious faculty of the mind, and it is the foundation of trope.

Now, when we use a word which has a great variety of uses, and can trace in this usage some one meaning as an attribute of consciousness, Hegel considers it to be the fundamental meaning, as shown by his practice. He affirms this, sometimes, when he says that every word must be taken in all its meaning, if it is logically used. The word comprehend is used as a sign of a mental and also of a physical act. I may say that I comprehend the pen, when I mean that I understand the pen, or I may say that the different parts of the pen are comprehended in one, as the pen itself. Now, in Hegel, the word for comprehend seems to have many meanings that are really comprehended in one, and, being an idealist, that one meaning is its psychic significance. There is no such thing as a pen with mechanical properties, but it exists only with the properties with which I endow it when I think it, for I create it with my thought. According to the Hegelian theory, it is nonsense to say that I think about a pen, but it is the "thing-in-itself," when I say I think the pen. This thing-in-itself is the noumenon of idealism. This knife is composed of the handle and its parts, and the blades with their parts, but, according to idealism, things are only what we think them to be, and the word composed, used in this manner, if properly understood, is but a psychic term, for, according to Kant, space is a form of thought, not of things. When we come properly to understand the world, that all things are thoughts, then we see that the real meaning of words is their psychic meaning, and that words can have but one meaning. As commonly understood, apprehended, composed, and embraced in the same senses, have synonymous meanings, but, according to Hegel, and to idealism generally since his time, synonymous words *always* have the same meaning, and that meaning must be found when it expresses a psychic fact. This is the secret of Hegel, and the key to his hieroglyphics, and, if consistently used to interpret the sayings of his logic, it becomes an open book. Now, when he uses a word for any property whatever, we must understand, if we follow Hegel in his argument, that the word is used in its psychic meaning. If we consistently carry out this rule, sentence by sentence, paragraph by paragraph, chapter by chapter, through the "Phenomenology," where it generally works, on through his "Logic," where, perhaps, it is the universal rule, we can translate his hieratic codex into demotic speech.

Permit a word of advice to the student who desires to accomplish this feat. First, read the works of Hegel's most devout disciples. Then take up Hegel himself. Then, after mastering Hegel, Kant's "Critique" will be an open book. The student must first learn the hieratic language, and then it is easy to read all of the works of the idealists.

Hegel accepted not only void space and void time as realities, but he accepted void essence and other nothings which he included under the term being, and sometimes under the term absolute. The world of sense is seen by every one to be a world of change, and he called it *becoming;* the fallacies, then, he called the *being*, or the *absolute*, and the realities the *becoming*. In his "Logic," he says:

FALLACIES OF IDEATION

"But this mere Being, as it is mere abstraction, is therefore the absolutely negative: which, in a similarly immediate aspect, is just Nothing.

"Hence was derived the second definition of the Absolute; the Absolute is the Nought. In fact this definition is implied in saying that the thing-in-itself is the indeterminate, utterly without form and so without content. . . .

"The proposition that Being and Nothing is the same seems so paradoxical to the imagination or understanding, that it is perhaps taken for a joke. And indeed it is one of the hardest things thought expects itself to do: for Being and Nothing exhibit the fundamental contrast in all its immediacy,—that is, without the one term being invested with any attribute which would involve its connexion with the other. This attribute, however, as the above paragraph points out, is implicit in them—the attribute which is just the same in both. So far the ˌdeduction of their unity is completely analytical: indeed the whole progress of philosophising in every case, if it be a methodical, that is to say, a necessary, progress, merely renders explicit what is implicit in a notion. It is as correct however to say that Being and Nothing are altogether different, as to assert their unity. The one is *not* what the other is. But since the distinction has not at this point assumed definite shape (Being and Nothing are still the immediate), it is, in the way that they have it, something unutterable, which we merely *mean.*"

What Hegel means is that the world of reality is the creation of the human mind out of nothing. Now, this creation of something out of nothing, as it produces the material universe, is kept in constant flux or change, for everything is in evolution and dissolution, and thus it is the becoming. While Spencer reifies the universe as force, and deems it the unknowable, Hegel reifies the universe as thought, and deems it the unutterable; so all metaphysical philosophers trace the universe into something occult.

Hegel attempts to forestall ridicule in the following language:

"No great expenditure of wit is needed to make fun of the maxim that Being and Nothing are the same, or rather to adduce absurdities which, it is erroneously asserted, are the consequences and illustrations of that maxim."

Then he goes on, by a method of logic which he calls dialectic, to show the validity of his proposition, in which he asserts:

"There is absolutely nothing whatever in which we cannot and must not point to contradictions or opposite attributes."

This logic is well worth perusal by the curious reader, as an example of mysterious arguments about mysteries, of propositions about the unutterable, of notions about the unknowable, and of attributes assigned to ghosts. In such manner, scholastic learning transmutes folk-lore into the semblance of wisdom, and the pathos of poetry. Lowell, with sympathetic love, has given fine expression to the thaumaturgy of transcendentalism, when he likens the gold-fish in the globe to souls imprisoned in the sphere of sense:

"Is it illusion? Dream-stuff? Show
Made of the wish to have it so?
'Twere something, even though this were all:
So the poor prisoner, on his wall
Long gazing, from the chance designs
Of crack, mould, weather-stain, refines
New and new pictures without cease,
Landscape, or saint, or altar-piece:
But these are Fancy's common brood,
Hatched in the nest of solitude;

This is Dame Wish's hourly trade,
By our rude sires a goddess made.

. . . .

"The worm, by trustful instinct led,
Draws from its womb a slender thread,
And drops, confiding that the breeze
Will waft it to unpastured trees;
So the brain spins itself, and so
Swings boldly off in hope to blow
Across some tree of knowledge, fair
With fruitage new, none else shall share:
Sated with wavering in the Void,
It backward climbs, so best employed,
And, where no proof is nor can be,
Seeks refuge with Analogy;
Truth's soft half-sister, she may tell
Where lurks, seld-sought, the other's well.

. . . .

"The things we see as shadows I
Know to be substance; tell me why
My visions, like those haunting you,
May not be as substantial too.
Alas, who ever answer heard
From fish, and dream-fish too? Absurd!
Your consciousness I half divine,
But you are wholly deaf to mine.
Go, I dismiss you; ye have done
All that ye could; our silk is spun;
Dive back into the deep of dreams,
Where what is real is what seems!
Yet I shall fancy till my grave
Your lives to mine a lesson gave;
If lesson none, an image, then,
Impeaching self-conceit in men
Who put their confidence alone
In what they call the Seen and Known."

Emerson sings of the mystery of transcendentalism:

"The Sphinx is drowsy,
　　Her wings are furled;
Her ear is heavy,
　　She broods on the world.
Who'll tell me my secret,
　　The ages have kept?
I awaited the seer,
　　While they slumbered and slept;—

"The fate of the man-child;
　　The meaning of man;
Known fruit of the unknown;
　　Daedalian plan;
Out of sleeping a waking,
　　Out of waking a sleep;
Life death overtaking;
　　Deep underneath deep?

　　. . . . . .

"Up rose the merry Sphinx,
　　And crouched no more in stone;
She melted into purple cloud,
　　She silvered in the moon;
She spired into a yellow flame;
　　She flowered in blossoms red;
She flowed into a foaming wave;
　　She stood Monadnoc's head."

Great are the poets of mysticism, but there is one greater:

"What a piece of work is man! How noble in reason! How infinite in faculties! in form, and moving, how express and admirable! in action, how like an angel! in apprehension, how like a god! the beauty of the world! the paragon of animals!"

CHAPTER XXV

SUMMARY

I have tried to demonstrate that an ultimate particle, and hence every body, has five essentials or concomitants, these terms being practically synonymous. It has been shown that there is something absolute and something relative in every one. The essentials of the particle are unity, extension, speed, persistence, and consciousness, which are absolute. The relations which arise from them, in order, are multeity, position, path, change, and choice, which give rise to number, extension, motion, time, and judgment, as properties that can be measured. It has been pointed out that particles are incorporated in bodies through affinity as choice, and by this incorporation the quantitative properties become classific properties which, in order, are class, form, force, causation, and conception. In the development of number into class, unity becomes kind and plurality becomes series. In the development of space into form, extension becomes figure and position becomes structure. In the development of motion into force, speed becomes velocity and path becomes inertia. In the development of time into causation, persistence becomes state and change becomes event. In the development of judgment into conception, consciousness becomes recollection and choice becomes inference.

As all particles, except those of the ether, are

organized into bodies, all of these bodies may be viewed or considered from two standpoints—internal and external. If we consider the body internally we consider its particles externally to one another; therefore, we are compelled to recognize the reciprocality of the two views—the quantitative view is equal to or the reciprocal of the classific view. Now, we have three terms, concomitancy, relativity, and reciprocality, which, in all science and especially in psychology, must clearly be distinguished. The failure to distinguish them creates the fog of metaphysic.

In the ether we do not know of the existence of bodies, but it seems probable that only particles exist. We do know of astronomical bodies, geonomic bodies, phytonomic bodies, zoönomic bodies, and demotic bodies. In the last class the particles do not lose their three degrees of freedom of motion, but this freedom is transmuted into coöperative reciprocality. The freedom of the particles by development of motility as a mode of motion becomes the self-activity of the individuals, which is exhibited in promoting the welfare of the individual and of the demotic body.

Properties are not creations of the mind; they are founded in nature and are recognized in nature in the plainest manner, hence they are not artificial, but natural. In molecules numbers are organized into kinds and series, that is, into classes, the kinds appearing as substances and the series as totalities of substances. In stars spaces are integrated and differentiated as figures and structures, and hence forms are primarily organized in stars. In geonomic bodies motions are organized as forces, being integrated and differentiated as coöperative spheres. In plants times are organized as causations, antecedent and consequent, as parents and children, and heredity thus appears. In animals judgments are organized, in which times become states as memories and changes as inferences. In this realm mind first appears as conception, for concepts require memory and inference, thus only animals have minds; plants do not have minds, but their particles have judgment, for particles have affinity and make judgments of association, and only such judgments. They do not have memory, nor do they have conception, therefore they do not have inference.

All particles of plants, rocks, and stars have judgments as consciousness and choice, but having no organization for the psychical functions they have not recollection and inference; they therefore do not have intellections or emotions. Only animal bodies have these psychical faculties. Molecules, stars, stones, and plants do not think; that which we have attributed to them as consciousness and choice is only the judgment of particles; but it is the ground, the foundation, the substrate of that which appears in animals when they are organized for conception.

That which perchance may be called hylozoism in this work must radically be distinguished from that hylozoism which appears in metaphysical speculation, when it attributes mind to inanimate bodies, or from that belief of early mythology which also attributes mind to inanimate things. It is this error of primeval savagery, called animism, from which civilized men have long ago logically revolted, that must be distinguished from the hylozoism herein propounded. Perhaps it is this repugnance to primeval error which has chiefly been instrumental in causing the rejection of the fundamental principles of concomitance in the science of mind, for it has occurred to great thinkers many times since the revival of science effected by Columbus and Copernicus.

It is marvelous how often it has occurred to the great thinkers of science as well as of metaphysics; but so far as I know it was never clearly formulated in such a manner as to become a scientific doctrine. It has been held that mind itself belongs to the inanimate realm, when it should have been held that consciousness and choice only are inherent in this realm, which is developed into psychic faculties only by the organization of animate bodies.

In these chapters it has been affirmed that every particle or body may be considered severally in its essentials or concomitants, and that if we consider one property and not the others we consider it abstractly. Abstraction, therefore, is the consideration of one property of a body, neglecting the others which we are compelled to posit.

We cannot conceive one property as existing independently of the others, but the discovery of one property leads the mind by a habit, which is inexorable, to postulate the others. This postulation of all properties from one, if neglected, leads to what has here been called reification. The mind that deals with things when it reasons, cannot make this mistake, but the mind that deals with words and thus reasons by the methods of scholastic logic, is liable to this error, for a particle or a body may be designated by the name of one of its properties. The failure to make this distinction may be called the ground of the failure of Aristotelian logic as distinguished from scientific logic.

Having set forth the reciprocal properties of bodies, a brief chapter is given to explain how properties become qualities, in which it is demonstrated that qualities arise through the consideration of properties in relation to the purposes of animal bodies, especially of human bodies.

The failure to distinguish between properties and qualities is the fundamental error of modern metaphysic. For twenty-five centuries many great thinkers have considered the properties of a body, which are founded upon its essentials, and which essentials are the thing-in-itself, as if they were qualities. Discovering that qualities are forever changing with the point of view, as the purpose of the individual is changed, the reality of properties was questioned.

The unreality of properties when they are confounded with qualities finds expression in many ways. Thus it is affirmed that man is the measure of things, or that man is the measure of qualities, meaning that things or their qualities are generated by the mind. This is true of qualities, as I use the term, but it is not true of properties. Still, the ancients retained sanity, and believed in the thing-in-itself, and called it a noumenon. An attribute of a thing which seems to vary with the point of view is called a phenomenon. Then, many properties are imperfectly cognized, and their explanation depends upon investigation which has come to be recognized as scientific research; hence properties that are still improperly explained are also called phenomena, but when better explained are called noumena. Thus noumenon as used by the ancients is a term which means the thing which changes with the point of view, whether it is a change of purpose or a change of explanation. Thus errors of cognition in properties are confounded with what I call qualities, and both are called phenomena.

An attempt is then made to demonstrate that the cognition of these properties gives rise to five psychic faculties, which we have called sensation, perception, apprehension, reflection, and ideation.

In developing the five faculties of intellection an endeavor has been made to set forth the nature of the judgment and to show that its validity depends upon verification. Repeated judgments from like sense impressions become habitual or intuitive. I here speak of habitual judgments of intellection as intuitive, as in a later work I shall speak of habitual judgments of emotion as instinctive, and consider presentative judgments as being inductive, and representative judgments as being deductive. The division of the faculties into sense perception, understanding, and reason, to which metaphysic has been committed in a more or less clearly defined manner, is here rejected as a schematization that leads to psychological confusion.

That speculation which deals with the properties of bodies as if they were qualities, I call metaphysic. That theorizing which distinguishes properties from qualities and deals with properties as realities, I call science. That speculation which fails to find consciousness as an essential or concomitant of bodies, but derives the mind from force or motion, I call materialism.

Metaphysic has a history which must be unraveled

to properly understand contemporaneous opinion at any one stage, but especially to understand the successive stages through which it has passed. Before the birth of chemistry man believed the elements to be earth, air, fire, and water, which elements were mixed in all of the bodies of the world, and it becomes necessary for us to understand how the attributes of bodies were assigned to the several elements. Not only was metaphysic founded upon these doctrines, but it was out of a philosophy of these elements that science itself was developed. Gradually in the history of civilization there grew up a doctrine of substance or substrate as something which is not one of the essentials of matter, as particle or body, but to which essentials adhere or inhere or subsist. This substrate or substance was supposed to be the vehicle of properties which emanate from it. Two relics of this doctrine are especially of interest to scientific men. It was long believed that heat and light are corpuscular, and that heat is given off from the substrate or substance of one body and taken up by another. Even Newton thought light to be corpuscular. The doctrine that motion as speed emanates from one body as a substance or substrate and passes to another, comes from this source. This relic of ancient philosophy clings to much of modern physics, and is the foundation of a body of speculation in which scientific men indulge when they theorize about the dissipation of motion, the exhaustion of the heat of the sun, and the general running down of the solar system into a state in which life will be impossible.

In a very brief and inadequate way I have tried to set forth the origin and history of fundamental fallacies relating to properties. This history commences with the early Greeks, but we cannot understand its origin without going back to an earlier stage of society than that in which history presents the philosophy of the Hellenic tribes.

In tribal society all honor is due to the progenitor of a tribe for his goodness and wisdom, and his commands have perpetual authority. The ancient time was the golden age; the present is a time of degeneracy. In tribal society to say that a man is elder is to say that he is wiser and better and must be obeyed. An ancient who lived in the ancient of days was supremely wise and good. He who can trace his ancestry farthest into antiquity has the most honorable beginning. The most ancient, the first, the progenitor, the prototype, is the one to whom all glory must be given. In savagery, authority, wisdom, honor, and parentage are so intimately associated in the minds of tribal men, that their demotic organization is dependent upon this compound concept, taken as a single principle. With these people demotic organization is founded upon the authority of the parent over the offspring. To be a parent is to have wisdom, and to be a parent is to have authority. The parent seems to have reason upon his side when he seeks to control the offspring, for the parent is the author of the offspring; therefore, the progenitor is the wise and the powerful, and this principle, which is at the foundation of tribal society, is so thoroughly interwoven into the habits of thought of the people, that it seems to them a self-evident proposition that the progenitor is wise and should rule.

When a group of kindred is considered with

parents and children, and collateral lines of consanguineal members, and further lines of kinship by affinity, the whole group organized into a tribe with authority in the relative elder, and all the items of authority parceled out in a hierarchy of real or conventional relative ages, we have the tribal plan of government.

Honor for ancestors is the most profound sentiment of savage men and is daily and systematically inculcated, so that the younger always yields obedience to the elder, and the elder is always held in reverence.

This principle leads to a gradation of the people of a savage tribe into recognized ranks by relative age, and if a man is promoted within a tribe, such promotion is a formal advancement in age, and kinship terms are readjusted so that the age received by promotion may be recognized in terms of address. In barbarism there comes another element to increase this respect, for the elder is not only obeyed, but is worshiped as a deity. In this manner often the chief of the gens, which is a group within the tribe, and also the chief of the tribe, is worshiped. Dead ancestors are also worshiped as ghosts. Clans of the savage tribe become gentes of the barbaric tribe, and the gentes are grouped in phratries as religious brotherhoods, and the dead chief of the phratry is usually worshiped, while other departed members of the phratry are also worshiped. Chiefs, who may be called the priests of the phratry when they become remarkable for their ability or for success in shamanism as diviners, medicine-men, and soothsayers, are held for a long time in great reverence, and their accomplishments are repeated in many a story. So in barbaric society the patriarch—the ancient—is held to be the progenitor or prototype of the gens, the phratry, or the tribe, as the case may be. In gentes, phratries, and tribes there is a constant veneration of ancestral ancients. This appears to have been the case among the Hellenic tribes, which belonged to that stage of culture which we call barbarism.

In savagery seven worlds are developed, as the east, west, north, south, zenith, nadir, and center; and they schematize or systematize all the attributes of things into seven groups. As geographic knowledge increases, those attributes which are assigned to the four quarters of the earth, are by natural methods transferred from the cardinal worlds to certain leading attributes of those worlds represented by earth, air, fire, and water. In this manner the worlds are transmuted into elements, but there still remain the zenith, nadir, and center—the zenith becoming a world of exalted attributes which they suppose to be good, the nadir becoming a world of evil.

Greek philosophy was developed at a time when tribal society was developing. Upon the ruins of tribal society imperialism was erected. The Greek philosophers inherited the cosmology of barbarism and with it the habits of thought characteristic of barbarism, especially the mental tendency to claim superiority for the ancient or first. Hence they claimed superiority for one or another of the four elements. Particularly was air, fire, or water held to be the first or progenitor of the others. In all their concepts about the absolutes of bodies, whether considering properties or qualities, there always seems to be a factor of this tribal concept. Thus we see that one of the barbaric elements was always taken as the substrate of the others. Thus was born the doctrine of substrate.

When imperialism had led to monotheism, and the school of theology was the school of philosophy also, a new substrate was discovered—the deity as something eminent in the world of attributes. At last, in comparatively modern times, another substrate was developed in speculation as a something to which the attributes could inhere. This reification still holds a place in the speculation even of scientific men and vitiates our popular physics. It is the chimera of substrate, this thing in itself as noumenon that leads to the belief in the world only as phenomenon. Since Berkeley and Hume a special school of metaphysicians has been developed who have the custody of this ghost and are its leal defenders. The fifth property, or consciousness as mind, is their ghost. These are the idealists. The war of philosophy is between Idealists and Materialists.

The philosophy here presented is neither Idealism nor Materialism; I would fain call it the Philosophy of Science.

INDEX

Activities, human, considered, 180.
Adaptation, laws of, 201.
Affinity, phenomena of, 41.
Animals, principles or properties of, 74.
Animals, environment of, 75.
Animals, motility in, 76.
Animals, heredity in, 76.
Animals, reproduction of, 77.
Animals, judgment in, 77.
Animals, systems of organs in, 78.
Animals, metabolic processes in, 79.
Animals, functions in, 83.
Animals, nervous system in, 87.
Animals, sense organs of, 89.
Apprehension, 237.
Apprehension, term restricted to judgment of force, 237.
Apprehension, both inductive and deductive, 243.
Apprehension, definition of, 250.
Apprehension, fallacies of, 352.
Assimilation, constructive, 66.
Assimilation, differentiating, 66.
Atmospheric agencies of disintegration, 48.

Berkeley, idealism of, 403.
Botany, facts relating to, 137.

Causation, primal fallacy of, 381.
Causation, study of, 186.
Cause and effect, 37.
Causes, genetic, 39.
Causes, teleologic, 39.
Chuar's illusion, 1.
Classification, definition of, 109.
Classification, method of, 113.
Classification, test of, 117.
Classification, goal of the science of, 118.
Classification, erroneous methods of, 119.
Classification, fundamental among Greeks, 122.
Classification, a tool of logic, 126.
Classification, logical and mathematical methods of, 131.
Cognition defined, 283.
Conception, a process of consolidating judgments, 214.
Consciousness considered, 211.
Coöperation discussed, 168.
Coöperation in celestial realm, 173.
Coöperation in terrestrial spheres, 174.
Coöperation in vegetal realm, 174.
Coöperation in zoönomic realm, 174.

TRUTH AND ERROR

Coöperation of systems with systems, 177.
Coöperation, societies formed by, 179.
Cosmology, origin of systems of, 377.
Culture, law of, 201.

Darwin, acceleration of evolution discovered by, 197.
Doctrines taught by modern science, 9.
Dynamics discussed, 152.

Earth, composition of, 42.
Earth, form of, 43.
Earth, geologic facts, 43.
Earth, aqueous envelope, 45.
Earth, oscillations of upheaval and subsidence, 45.
Earth, structure of crust of, 47.
Earth, changes effected in crust by water, 60.
Effort, law of, 200.
Environment, effect of, 204.
Environment of animals, 75.
Evolution discussed, 183.
Evolution, primal law of, 188.
Evolution, organization of demotic life a factor in, 200.

Force, compound of motions, 36.
Forces, outline of, 171.
Functions, control of, 178.

Generations or properties of plants, 64.
Geochemism, the fundamental energy, 59.
Geonomic bodies, properties of, 42.
Geonomic realm, constitution of, 192.
Geonomy, divisions of, 136.

Hallucinations defined, 313.
Hallucinations, so-called census of, 315.
Hallucinations, classification of, 320.

Hallucinations, among North American Indians, 330.
Hegel, fallacies of, 405.
Heredity, effect of, 177.
Heredity, element of, in plants, 65.
Heredity, law of, 199.
Heterogeneity, law of, 199.
Homologies, extended from atom to organism, 145.
Homologies, hierarchy of, throughout universe, 146.
Homologies, illustrated in organization of human society, 146.
Homologies, in natural organization, 147.
Homology discussed, 133.
Human body, a hierarchy of conscious bodies, 86.
Hylozoism, theory of, 95.
Hypotheses relating to changes in earth's crust, 51.

Ideation discussed, 264.
Ideation, fallacies of, 391.
Inertia, definition of, 360.
Intellections discussed, 278.
Intellections, psychology a system of, 278.
Intellections, faculties of, 279.

Judgment, psychic elements of, 280.
Judgment relating to cause and effect, 282.

Lamarck, motile state of matter discovered by, 199.
La Place, on genesis of heavenly bodies, 190.
Law governing phenomena of earth's crust, search for, 50.
Locke, John, on accidents, 401.

Mathematics of motion, science of, 27.
Matter, definition of, 12.
Memory, physiological conception of, 333.
Metabolism in plants, 72.

Metabolism in animals, 74.
Metagenesis, a process of causality, 63.
Metamorphoses of mineral substances, 56.
Metaphysical reasoning, errors of, 276.
Metaphysisis, a succession of changes of force, 59.
Mill, John Stuart, on causation, 38.
Misperceptions, illustrations of, 338-339.
Molar bodies, definition of, 21, 129.
Molecular bodies, internal relations of, 21.
Morphology, illustrated by animals and their organs, 142.
Morphology, illustrated by insects, 145.
Morphology and classification, relation between, 148.
Morphology of plant phytons, 70.
Motion of atoms, 152.
Motion, Descartes' theory, 153.
Motion, laws of, 163.
Motion, vibratory and structural, 169.
Motion, persistence of, 358.
Motion, concepts of, 361.
Myths, discussion of, 381.

Nature, five fundamental realms of, 96.
Newton, theory of inertia, 161.

Ontology, philosophy of, 399.

Parish, on hallucinations and illusions, 311.
Particles, inanimate, essentials of, 16.
Particles of matter, affinity of, 31.
Particles of matter, combination into molecules, 31.
Particles of matter, classification, 31.
Particles of matter, vibration of, 35.
Particles of matter, persistence of, interrupted, 36.
Perception discussed, 226.
Perception, fallacies of, 335.
Perception, as a mental phenomenon, 341.
Perception, the interpretation of a symbol, 342.
Phenomena, erroneously classified by Mill, 120.
Philosophy, transcendental, errors of, 149.
Philosophy of the unknowable, refutation of Spencer's, 370.
Plants, chemical laboratories, 69.
Plants, cells, tissues and forms of, 69.
Plants, conditions of life, 72.
Processes or the properties of geonomic bodies, 42.
Properties, essentials of, 9.
Protoplasm, constitution of, 64.
Psychophysics, science of, 120.

Qualities, definition of, 98.
Qualities, distinct from properties, 100.
Qualities, errors of Locke in relation to, 102.
Qualities, termed attributes by Spencer, 105.
Qualities, Berkeley's opinions in relation to, 107.
Qualities, Hume, Kant, Schilling and Fichte on, 107.
Quantities or properties that are measured, 20.

Reflection discussed, 251.
Reflection, concepts of, 253.
Reflection, definition of, 263.
Reflection, process of combining judgments by, 289.
Reflection, fallacies of, 374.
Reification, origin of, 3.
Relations that must exist between particles, 23.

Science and speculation, issue between, 150.
Science and metaphysics, difference of methods, 184.
Sciences, distinction between classific and quantitative, 246.
Scientific research, definition of, 7.
Sensation discussed, 207.
Sensation and feeling, difference between, 207.
Sensation, fallacies of, 307.
Sense impressions, 223.
Senses, vicarious feelings, 209.
Signatures, the doctrine of, 385.

Solar system, motions within, 34.
Space, definition of, 133.
Specters, classification of, 348.
Structural geology, 56.
Substrates, recapitulation of, 222.
Survival, law of, 199.
Symbolism, explanation of the laws of, 300.

Time, development by incorporation, 36.
Transmutation of substances in rocks, 54.

Verification, methods of, 220.
Volcanic eruptions, 47.

The Religion of Science Library.

A collection of bi-monthly publications, most of which are reprints of books published by The Open Court Publishing Company. Yearly, $1.50 Separate copies according to prices quoted. The books are printed upon good paper, from large type.

The Religion of Science Library, by its extraordinarily reasonable price, will place a large number of valuable books within the reach of all readers
The following have already appeared in the series:

No. 1. *The Religion of Science.* By PAUL CARUS. 25c.
2. *Three Introductory Lectures on the Science of Thought.* By F. MAX MÜLLER. 25c.
3. *Three Lectures on the Science of Language.* By F. MAX MÜLLER. 25c.
4. *The Diseases of Personality.* By TH. RIBOT. 25c.
5. *The Psychology of Attention.* By TH. RIBOT. 25c.
6. *The Psychic Life of Micro-Organisms.* By ALFRED BINET. 25c.
7. *The Nature of the State.* By PAUL CARUS. 15c.
8. *On Double Consciousness.* By ALFRED BINET. 15c.
9. *Fundamental Problems.* By PAUL CARUS. 50c.
10. *The Diseases of the Will.* By TH. RIBOT. 25c.
11. *The Origin of Language.* By LUDWIG NOIRÉ. 15c.
12. *The Free Trade Struggle in England.* By M. M. TRUMBULL. 25c.
13. *Wheelbarrow on the Labor Question.* By M. M. TRUMBULL. 35c.
14. *The Gospel of Buddha.* By PAUL CARUS. 35c.
15. *The Primer of Philosophy.* By PAUL CARUS. 25c.
16. *On Memory,* and *The Specific Energies of the Nervous System.* By PROF EWALD HERING. 15c.
17. *The Redemption of the Brahman.* A Tale of Hindu Life. By RICHARD GARBE. 25c.
18. *An Examination of Weismannism.* By G. J. ROMANES. 35c.
19. *On Germinal Selection.* By AUGUST WEISMANN. 25c.
20. *Lovers Three Thousand Years Ago.* By T. A. GOODWIN. 15c.
21. *Popular Scientific Lectures.* By ERNEST MACH. 50c.
22. *Ancient India : Its Language and Religions.* By H. OLDENBERG. 25c
23. *The Prophets of Ancient Israel.* By PROF. C. H. CORNILL. 25c.
24. *Homilies of Science.* By PAUL CARUS. 35c.
25. *Thoughts on Religion.* By G. J. ROMANES. 50 cents.
26. *The Philosophy of Ancient India.* By PROF. RICHARD GARBE.
27. *Martin Luther.* By GUSTAV FREYTAG. 25c.
28. *English Secularism.* By GEORGE JACOB HOLYOAKE. 25c.
29. *On Orthogenesis.* By TH. EIMER. 25c.
30. *Chinese Philosophy.* By PAUL CARUS. 25c.
31. *The Lost Manuscript.* By GUSTAV FREYTAG. 60c.
32. *A Mechanico-Physiological Theory of Organic Evolution.* By CARL VON NAEGELI. 15c.

THE OPEN COURT PUBLISHING CO.
324 DEARBORN STREET, CHICAGO, ILL.
LONDON: Kegan Paul, Trench, Trübner & Co.

CATALOGUE OF PUBLICATIONS

OF THE

OPEN COURT PUBLISHING CO.

COPE, E. D.
THE PRIMARY FACTORS OF ORGANIC EVOLUTION.
121 cuts. Pp., xvi, 547. Cloth, $2.00, net.

MÜLLER, F. MAX.
THREE INTRODUCTORY LECTURES ON THE SCIENCE OF THOUGHT.
128 pages. Cloth, 75 cents. Paper, 25 cents.
THREE LECTURES ON THE SCIENCE OF LANGUAGE.
112 pages. 2nd Edition. Cloth, 75 cents. Paper, 25c.

ROMANES, GEORGE JOHN.
DARWIN AND AFTER DARWIN.
An Exposition of the Darwinian Theory and a Discussion of Post-Darwinian Questions. Three Vols., $4.00. Singly, as follows:
1. THE DARWINIAN THEORY. 460 pages. 125 illustrations. Cloth, $2.00.
2. POST-DARWINIAN QUESTIONS. Heredity and Utility. Pp. 338. $1.50.
3. POST-DARWINIAN QUESTIONS. Isolation and Physiological Selection. Pp. 181. $1.00.
AN EXAMINATION OF WEISMANNISM.
236 pages. Cloth, $1.00. Paper, 35c.
THOUGHTS ON RELIGION.
Edited by Charles Gore, M. A., Canon of Westminster. Third Edition, Pages, 184. Cloth, gilt top, $1.25.

RIBOT, TH.
THE PSYCHOLOGY OF ATTENTION.
THE DISEASES OF PERSONALITY.
THE DISEASES OF THE WILL.
Authorised translations. Cloth, 75 cents each. Paper, 25 cents. *Full set, cloth, $1.75, net.*
EVOLUTION OF GENERAL IDEAS.
(In preparation.)

MACH, ERNST.
THE SCIENCE OF MECHANICS.
A CRITICAL AND HISTORICAL EXPOSITION OF ITS PRINCIPLES. Translated by T. J. McCORMACK. 250 cuts. 534 pages. ½ m., gilt top. $2.50.
POPULAR SCIENTIFIC LECTURES.
Third Edition. 415 pages. 59 cuts. Cloth, gilt top. Net, $1.50.
THE ANALYSIS OF THE SENSATIONS.
Pp. 208. 37 cuts. Cloth, $1.25, net.

LAGRANGE, J. L.
LECTURES ON ELEMENTARY MATHEMATICS.
With portrait of the author. Pp. 172. Price, $1 00, net.

HUC AND GABET, MM.
TRAVELS IN TARTARY, THIBET AND CHINA.
(1844-1846.) Translated from the French by W. Hazlitt. Illustrated with 100 engravings on wood. 2 vols. Pp. 28 + 660. Cl., $2.00 (10s.).

CORNILL, CARL HEINRICH.
THE PROPHETS OF ISRAEL.
Popular Sketches from Old Testament History. Pp., 200. Cloth, $1.00.
HISTORY OF THE PEOPLE OF ISRAEL.
Pp. vi + 325. Cloth, $1.50.

BINET, ALFRED.
THE PSYCHIC LIFE OF MICRO-ORGANISMS.
Authorised translation. 135 pages. Cloth, 75 cents; Paper, 25 cents.
ON DOUBLE CONSCIOUSNESS. See No. 8, Relig. of Science Library.

WAGNER, RICHARD.
   A PILGRIMAGE TO BEETHOVEN.
      A Novelette. Frontispiece, portrait of Beethoven. Pp. 40. Boards, 50c

HUTCHINSON, WOODS.
   THE GOSPEL ACCORDING TO DARWIN.
      Pp., xii + 241. Price, $1.50.

FREYTAG, GUSTAV.
   THE LOST MANUSCRIPT. A Novel.
      2 vols. 953 pages. Extra cloth, $4.00. One vol., cl., $1.00; paper, 75c
   MARTIN LUTHER.
      Illustrated. Pp. 130. Cloth, $1.00. Paper, 25c.

TRUMBULL, M. M.
   THE FREE TRADE STRUGGLE IN ENGLAND.
      Second Edition. 296 pages. Cloth, 75 cents; paper, 25 cents.
   WHEELBARROW: ARTICLES AND DISCUSSIONS ON THE LABOR QUESTION
      With portrait of the author. 303 pages. Cloth, $1.00; paper, 35 cents.

GOETHE AND SCHILLER'S XENIONS.
   Selected and translated by Paul Carus. Album form. Pp., 162. Cl., $1.00

OLDENBERG, H.
   ANCIENT INDIA: ITS LANGUAGE AND RELIGIONS.
      Pp. 100. Cloth, 50c. Paper, 25c.

CARUS, PAUL.
   THE ETHICAL PROBLEM.
      90 pages. Cloth, 50 cents; Paper, 30 cents.
   FUNDAMENTAL PROBLEMS.
      Second edition, enlarged and revised. 372 pp. Cl., $1.50. Paper, 50c.
   HOMILIES OF SCIENCE.
      317 pages. Cloth, Gilt Top, $1.50.
   THE IDEA OF GOD.
      Fourth edition. 32 pages. Paper, 15 cents.
   THE SOUL OF MAN.
      With 152 cuts and diagrams. 458 pages. Cloth, $3.00.
   TRUTH IN FICTION. TWELVE TALES WITH A MORAL.
      Fine laid paper, white and gold binding, gilt edges. Pp. 111. $1.00.
   THE RELIGION OF SCIENCE.
      Second, extra edition. Price, 50 cents. R. S. L. edition, 25c. Pp. 103.
   PRIMER OF PHILOSOPHY.
      240 pages. Second Edition. Cloth, $1.00. Paper, 25c.
   THE GOSPEL OF BUDDHA. According to Old Records.
      4th Edition. Pp., 275. Cloth, $1.00. Paper, 35 cents. In German, $1.25.
   BUDDHISM AND ITS CHRISTIAN CRITICS.
      Pages, 311. Cloth, $1.25.
   KARMA. A STORY OF EARLY BUDDHISM.
      Illustrated by Japanese artists. 2nd Edition. Crêpe paper, 75 cents.
   NIRVANA: A STORY OF BUDDHIST PSYCHOLOGY.
      Japanese edition, like *Karma*. $1.00.
   LAO-TZE'S TAO-TEH-KING.
      Chinese-English. With introduction, transliteration, Notes, etc. Pp 360. Cloth, $3.00.

GARBE, RICHARD.
   THE REDEMPTION OF THE BRAHMAN. A TALE OF HINDU LIFE.
      Laid paper. Gilt top. 96 pages. Price, 75c. Paper, 25c.
   THE PHILOSOPHY OF ANCIENT INDIA.
      Pp. 89. Cloth, 50c. Paper, 25c.

HUEPPE, FERDINAND.
   THE PRINCIPLES OF BACTERIOLOGY.
      28 Woodcuts. Pp., 350+. Price, $1.75. (In preparation.

THE OPEN COURT

A MONTHLY MAGAZINE

Devoted to the Science of Religion, the Religion of Science, and the Extension of the Religious Parliament Idea.

THE OPEN COURT does not understand by religion any creed or dogmatic belief, but man's world-conception in so far as it regulates his conduct.

The old dogmatic conception of Religion is based upon the science of past ages; to base religion upon the maturest and truest thought of the present time is the object of *The Open Court*. Thus, the religion of *The Open Court* is the Religion of Science, that is, the religion of verified and verifiable truth.

Although opposed to irrational orthodoxy and narrow bigotry, *The Open Court* does not attack the properly religious element of the various religions. It criticises their errors unflinchingly but without animosity, and endeavors to preserve of them all that is true and good.

The current numbers of *The Open Court* contain valuable original articles from the pens of distinguished thinkers. Accurate and authorized translations are made in Philosophy, Science, and Criticism from the literature of Continental Europe, and reviews of noteworthy recent investigations are presented. Portraits of eminent philosophers and scientists are published, and appropriate illustrations accompany some of the articles.

Terms: $1.00 a year; $1.35 to foreign countries in the Postal Union. Single Copies, 10 cents.

THE MONIST

A QUARTERLY MAGAZINE OF

PHILOSOPHY AND SCIENCE.

THE MONIST discusses the fundamental problems of Philosophy in their practical relations to the religious, ethical, and sociological questions of the day. The following have contributed to its columns:

Prof. Joseph Le Conte,	Prof. G. J. Romanes,	Prof. C. Lombroso,
Dr. W. T. Harris,	Prof. C. Lloyd Morgan,	Prof. E. Haeckel,
M. D. Conway,	James Sully,	Prof. H. Höffding,
Charles S. Peirce,	B. Bosanquet,	Dr. F. Oswald,
Prof. F. Max Müller,	Dr. A. Binet,	Prof. J. Delbœuf,
Prof. E. D. Cope,	Prof. Ernst Mach,	Prof. F. Jodl,
Carus Sterne,	Rabbi Emil Hirsch,	Prof. H. M. Stanley,
Mrs. C. Ladd Franklin,	Lester F. Ward,	G. Ferrero,
Prof. Max Verworn,	Prof. H. Schubert,	J. Venn,
Prof. Felix Klein,	Dr. Edm. Montgomery,	Prof. H. von Holst,

Per copy, 50 cents; Yearly, $2.00. In England and all countries in U.P.U. per copy, 2s 6d; Yearly, 9s 6d.

CHICAGO:

THE OPEN COURT PUBLISHING CO.,

Monon Building, 324 Dearborn St.

LONDON: Kegan Paul, Trench, Trübner & Co.

www.ingramcontent.com/pod-product-compliance
Lightning Source LLC
Chambersburg PA
CBHW020533300426
44111CB00008B/642